T0185650

Introduction to Bayesian Methods in Ecology and Natural Resources

Edwin J. Green · Andrew O. Finley ·
William E. Strawderman

Introduction to Bayesian Methods in Ecology and Natural Resources

 Springer

Edwin J. Green
Department of Ecology
Evolution and Natural Resources
Cook College, Rutgers University
New Brunswick, NJ, USA

Andrew O. Finley
Department Forestry & Geography
Michigan State University
East Lansing, MI, USA

William E. Strawderman
Department of Statistics
Rutgers University
Piscataway, NJ, USA

ISBN 978-3-030-60752-4   ISBN 978-3-030-60750-0   (eBook)
https://doi.org/10.1007/978-3-030-60750-0

This Springer imprint is published by the registered company Springer Nature Switzerland AG
The registered company address is: Gewerbestrasse 11, 6330 Cham, Switzerland

To

Rosemary, Jimmy, Tara, Anthony, Marla, EJ, Mandy, Jane, and Amelia

Sarah, Ava, Oliver, Callum, James, Linda, and Gail

Rob, Myla, Bill, Jinny, Heather, Jim, Kay, Matt, Will, Tom, Evan, Emma, AJ, and Lily

In memory of Susan

Contents

List of Code Boxes

---

[1]https://github.com/finleya/GFS

Chapter 1
Introduction

Bayesian inference in the sciences has become remarkably widespread in the wake of the Markov chain Monte Carlo (MCMC) revolution of the 1990s. MCMC methods permit solutions to Bayesian problems which had previously been mathematically intractable. MCMC methods have simplified Bayesian inference to the point where it is often arguably simpler than conventional statistical approaches. However, ease of use is not and should not be a compelling reason on its own to justify a statistical approach. Hence in this text we will endeavor to motivate Bayesian analyses in ecological and/or natural resource management problems on the grounds that these methods readily permit scientists to model phenomena of interest in realistic ways. One does not need to accept the view that Bayesian methods are philosophically more appealing than conventional methods in order to use them effectively. We suspect that many scientists use Bayesian methods today only because they perform well and allow the scientist to directly answer the specific question posed.

There has been a long and often contentious debate in the literature regarding the relative merits of Bayesian and non-Bayesian statistical methods. To date, the debate has not been conclusively settled and perhaps it never will be. Even the authors of this text do not completely agree on some issues. However, given their increasingly widespread use, it is apparent that the modern scientist must at least understand Bayesian methods and preferably have them readily available in their toolbox. To that end, our goal in writing this text is to present some common Bayesian data analysis methods in a manner that will be understandable and readily available to students and scientists in the various fields of Ecology and Natural Resource Management.

We assume that the reader is a typical graduate student in Ecology and/or Natural Resource Management. In our experience, such a student has had some training in Analytic Geometry and Calculus and one or two undergraduate courses in Statistics. We attempt to explain concepts and ideas in a way that will be accessible to such students. After reading this book, a student should be able to pursue Bayesian analyses and read more sophisticated texts on the subject.

Modern Bayesian inference relies heavily on computer simulation. To relieve scientists of the burden of writing new code for every problem, a number of Bayesian computing packages have arisen. Chief among these is BUGS (Bayesian inference

© Springer Nature Switzerland AG 2020
E. J. Green et al., *Introduction to Bayesian Methods in Ecology and Natural Resources*,
https://doi.org/10.1007/978-3-030-60750-0_1

Using Gibbs Sampling), and its successors, WinBUGS and OpenBUGS (Lunn et al. 2000). These packages have the virtue of being freely available from the BUGS project (links are provided in Appendix C). Other widely used packages include JAGS (Just Another Gibbs Sampler (Plummer 2003), Stan (Carpenter et al. 2017), and NIMBLE (de Valpine et al. 2017). All examples in this book (other than those in Chap. 8: Spatial Models) are performed using OpenBUGS, and the code is provided in text boxes. A short tutorial on OpenBUGS is presented in Appendix C.

In Chap. 2 we cover various theoretical probability distributions and densities. For that purpose, we employ the freely available statistical computing package R, and we present the requisite R code. R is available from the Comprehensive R Archive Network (CRAN, www.cran.r-project.org). Users may also find tutorials and links to user's guides at CRAN. We make no attempt to instruct the reader on the use of R.

We have included many code boxes with either R or OpenBUGS code to help the reader perform analyses or produce graphs. Many of the R code boxes entail reading in `.txt` files produced by the `coda` option in OpenBUGS. In such cases, we indicate that the user must set the working directory to that where the `.txt` file is found. Other options (such as the size of the joint posterior sample generated in OpenBUGS or the order of the variables in the output files) depend on the parameters used in the OpenBUGS program. It is up to the user to ensure that these are specified correctly.

All the code in the code boxes (and larger data sets referenced in some boxes and exercises) is available online at https://github.com/finleya/GFS. Rather than repeating this lengthy URL every time it is needed, we will indicate that a data set or piece of code is available `online`.

1.1 Bayesian and Non-Bayesian Inference

This book is concerned with making inferences from data. Although the techniques we describe are useful in many different disciplines, we focus on data which arise in forestry, ecology, and wildlife biology. Even within that small slice of scientific disciplines, the interests and orientations of scientists are remarkably varied; hence we will refer to the reader generically as a "scientist." In our view, a scientist is interested in statistical procedures insofar as they permit legitimate inferences about populations from sample data. These inferences may take on various forms, such as hypothesis testing or model development. Most scientists are aware that, broadly speaking, there are at least two classes of statistics: Bayesian and non-Bayesian. The latter class is often referred to as "classical" or "frequentist" but for our purposes the latter class can be considered to be any inferential system not based on Bayes theorem. Many scientists have probably noted an increase in the use of Bayesian procedures since the early 1990s and may wonder what caused this increase, and what the big fuss is all about. In this Chapter we introduce Bayesian inference, briefly touch on its history and the controversy over its use, and conclude with a short discussion of the some of the reason(s) behind its increase in popularity.

1.2   Bayes Theorem

Let $P(\cdot)$ denote the probability of the quantity inside the parentheses. It may surprise some scientists to learn that there is no universally accepted definition of probability. Kolmogorov (1933) set out a set of three axioms that any coherent system of probability should satisfy, but the axioms do not define a probability system. Broadly speaking, there are two common notions of probability: frequentism and personal probability. In frequentism, the probability of an event occurring is defined as the percentage of times it occurs in a long series of trials. Unfortunately, the definition does not define how long "long" is. Also, it is not helpful in answering questions like "What is the probability of life on Mars?", where it is difficult to imagine repeated trials. On the other hand, personal probability, or subjective probability, is what the scientist believes in their mind and may vary among individuals, depending on that individual's background and knowledge (e.g., see de Finetti 1974). In general, the authors favor the personal probability approach, yet as will become evident in the subsequent chapters, like most modern Bayesians we make liberal use of vague or noninformative priors to avoid the burden of constructing subjective prior probability distributions for every problem.

Suppose we are considering two arbitrary events: **A** and **B**. Consider the probability that both **A** and **B** occur; this is the intersection of **A** and **B**, and in introductory probability texts it is usually written as $P(\mathbf{A} \cap \mathbf{B})$. In statistical work, it is customary to suppress the intersection symbol, and write the probability of the intersection of **A** and **B** as $P(\mathbf{A}, \mathbf{B})$. Now, suppose we wish to know the probability that event **B** will occur, given that event **A** has already occurred. We write this as $P(\mathbf{B} \mid \mathbf{A})$. This is the *probability of* **B**, *given* **A**, or the *probability of* **B**, *conditional on* **A**. If the events **A** and **B** are independent, then $P(\mathbf{B} \mid \mathbf{A}) = P(\mathbf{B})$, i.e., the outcome of **A** tells us nothing about the probabilities of the outcomes of **B**.

The multiplicative rule of probability (see, e.g., Harris 1966, p. 11) states

$$P(\mathbf{A}, \mathbf{B}) = P(\mathbf{A})P(\mathbf{B} \mid \mathbf{A}). \tag{1.2.1}$$

As Berry (1997) reports, Eq. 1.2.1 is intuitive. Berry asks the reader to consider the probability of observing two aces in two random drawings from a deck of cards, assuming sampling without replacement, i.e., that the first card drawn is not returned to the deck prior to drawing the second card. Most people would start by saying "first we need to compute the probability of an ace on the first draw; since there are 52 cards and four aces, the probability of this is 4/52. Then, we need the probability of an ace on the second draw; the probability of this is 3/51, because there are only three aces among the remaining 51 cards. Hence probability of observing two aces on two draws is 4/52 times 3/51." These people have just used Eq. 1.2.1.

Re-arranging terms in Eq. 1.2.1, we find

$$P(\mathbf{B} \mid \mathbf{A}) = \frac{P(\mathbf{A}, \mathbf{B})}{P(\mathbf{A})}. \tag{1.2.2}$$

Observe that if P(**A**) = 0, then P(**B** | **A**) is undefined, as it should be since P(**A**) = 0 means the event **A** is impossible, and it is pointless to consider the probability of **B** given an impossible event.

Now, since the events **A** and **B** were arbitrarily defined, we also have

$$P(\mathbf{A}, \mathbf{B}) = P(\mathbf{B}) P(\mathbf{A} \mid \mathbf{B}). \tag{1.2.3}$$

Replacing the numerator on the right-hand side (RHS) of (1.2.2) with the RHS of (1.2.3) yields

$$P(\mathbf{B} \mid \mathbf{A}) = \frac{P(\mathbf{B}) P(\mathbf{A} \mid \mathbf{B})}{P(\mathbf{A})}. \tag{1.2.4}$$

Equation 1.2.4 is the celebrated Bayes theorem. It reveals the proper, *and only*, way to "reverse" the conditioning of a probability statement; on the RHS we have event **A** conditioned on event **B**, while on the left-hand side (LHS) we have the reverse. It is important to note that this form of Bayes theorem is non-controversial and is an elementary result of the rule of multiplicative probability (Eqs. 1.2.1 and 1.2.3). It has many straightforward uses in this form, such as in image classification (Green et al. 1992; Richards and Jia 2006) or clinical diagnostic testing (Joseph et al. 1995; Spiegelhalter et al. 1999). The fun begins when Bayes theorem is used as a basis for scientific inference.[1]

1.3 Bayesian Inference

Suppose we collect sample data y on some variable, say Y. Further suppose that we believe the sampling distribution of Y (i.e., the distribution from which the observations on Y arise) is indexed, or governed, by some unknown parameter(s) θ. For convenience, in this section we will discuss θ as if it was one-dimensional, i.e., a scalar, however the reader should be aware that θ is frequently multi-dimensional. Assume we are interested in making inferences about the parameter θ based on the sample data y. Now, since we know the sampling distribution, we can evaluate $P(y \mid \theta)$, the probability of y conditioned on θ, for any value of θ. But that's not what we really want to know; we don't know θ, we only know y. It seems self-evident to seek the probability distribution of what we *don't* know (θ), conditioned on what we *do* know (y). Application of Bayes theorem yields

---

[1] Bayes theorem is named after the 18th Century English cleric Thomas Bayes (c. 1702–1761). The theorem was derived in an essay published posthumously by his friend, Richard Price. As noted by Bernardo and Smith (1994), we don't know how Rev. Bayes would feel about the system of inference attributed to him.

$$P(\theta \mid y) = \frac{P(\theta)P(y \mid \theta)}{P(y)}. \tag{1.3.1}$$

Equation 1.3.1 is the form of Bayes theorem used for scientific inference. Recall that we are interested in making inferences about θ. Since the denominator on the RHS of (1.3.1), $P(y)$, is independent of θ, we can learn nothing about θ from this term. Furthermore, once y is observed, $P(y)$ is fully specified and has a fixed value, say c. So, we can rewrite (1.3.1) as

$$P(\theta \mid y) = \frac{P(y \mid \theta)P(\theta)}{c}, \tag{1.3.2}$$

$$\propto P(y \mid \theta)P(\theta). \tag{1.3.3}$$

In Eq. 1.3.2, c^{-1} is a normalizing constant; its function is to ensure that the total probability sums (or integrates) to 1. The first term on the RHS of expression (1.3.3) is the probability of y given the parameter θ. In non-Bayesian as well as Bayesian statistics, following Fisher (1922) it has become usual to consider this as a function of θ rather than of y. When viewed in this way, $P(y \mid \theta)$ is called the likelihood function of θ given y, and is written as $L(\theta \mid y)$.[2] The value of θ which maximizes $L(\theta \mid y)$ is called the maximum likelihood estimate (e.g., see Casella and Berger 2001). If we adopt the likelihood notation, then we can re-write expression (1.3.3) as

$$P(\theta \mid y) \propto L(\theta \mid y)P(\theta). \tag{1.3.4}$$

Expression (1.3.4) is the form usually used in reports on Bayesian analyses. As mentioned above, $L(\theta \mid y)$ is widely used in non-Bayesian as well Bayesian inference and is not controversial. In Bayesian inference, it is often instructive to think of this as the sampling distribution for y, i.e., a mathematical description of the process in Nature that generates the data we observe.

The second term on the RHS of (1.3.4), $P(\theta)$, is called the prior distribution of θ; it represents what was known (or believed) about θ before the data y were observed. The term on the LHS of (1.3.4), $P(\theta \mid y)$, is called the posterior distribution of θ; it synthesizes all that was known about θ before the data were observed plus what was learned about θ from the data. In the Bayesian paradigm, all inferences on θ derive from the posterior distribution.

The controversy over Bayesian inference arises primarily over the term $P(\theta)$, and basically boils down to whether or not it is admissible to place a distribution on θ. Non-Bayesians make a distinction between random variables and population parameters. In this view, the parameter θ is fixed; we just don't know what it is. In many situations we can "imagine" measuring all the individuals in a population and then computing the exact value of θ, even though it might be prohibitively expensive or inefficient to do so. *If* we knew the values of all the elements in the population, we could calculate the true value of θ. Hence, since θ is fixed, it would be incorrect

---

[2]Some authors use $L(\theta \mid y)$ to indicate the *log* of the likelihood function.

to place a probability distribution on it. For example, suppose we know that the mean height of a specific population of trees is exactly 14 m (for the moment let's not concern ourselves with how we could possibly know this). Then the probability that the mean height is 14 m is 1.0 and the probability that it is any other value is 0. Thus this distribution has a mass of 1 at a single point: 14. Such a distribution is degenerate, and not really a distribution at all. Hence in the non-Bayesian view it is not permissible to construct probability distributions for parameters. They are fixed constants; the probability that they equal their true value is 1.0 and the probability that they assume any other value is 0.

Bayesians view the world differently. To them, the distinction between random variables and parameters is artificial and largely irrelevant. Bayesians are also interested in two classes of objects, but instead of random variables and parameters, they are interested in what is known and what is unknown. In the Bayesian view, probability distributions are used for expressing the state of our knowledge about any unknown object. Hence, since θ is generally unknown, Bayesians find it perfectly acceptable to place a probability distribution on it.

Interestingly, Bayesian texts often contain a summary of the differences between Bayesian and non-Bayesian inference (and usually a vigorous defense of the Bayesian view). For example, see Berger (1985), Robert (2001), Bernardo and Smith (1994), Gelman et al. (2013), Carlin and Louis (2009), or the classic Jeffreys (1935). On the other hand, texts on non-Bayesian inference often do not devote much space, if any, to Bayesian inference. A good historical account of Bayesian inference is contained in Stigler (1986), and an excellent non-mathematical treatment of the history may be found in McGrayne (2012). Readers interested in the controversy between the Bayesian and non-Bayesian view are referred to the above mentioned texts, and to the discussion following the classic papers by Lindley (1990), Lindley and Phillips (1976), Lindley and Smith (1972), and the references contained therein. A vast literature on the relative merits of Bayesian and non-Bayesian inference exists, and we cannot reproduce all the arguments here. We do however feel that it is necessary to cover the salient points so that the reader can make up their own mind. Bear in mind that although we (the authors) endeavor to summarize the arguments fairly, we do have reasonably firm opinions regarding the merits of Bayesian inference, so we cannot be entirely objective.

1.4   Pros and Cons of Bayesian Inference

Prior to about 1990, there was a practical objection to the use of Bayesian inference; in many cases it was impossible to solve for the constant c in (1.3.2), and hence it was difficult or impossible to solve for the posterior distribution; i.e., the analysis often could not be done, or if it could it required great skill in numerical analysis. The MCMC revolution of the early 1990s largely eliminated this concern and has made Bayesian analysis almost routinely possible now.

Current objection(s) to Bayesian inference continue to revolve around the specification of prior distributions for population parameters. Since priors must be specified before data are observed, they are clearly not dependent on the data and hence other factors besides the sample data may influence the analysis. This is in stark contrast with the usual stated goal of objectivity in scientific analyses. Furthermore, if two scientists analyze the same data but employ different prior distributions, they may or may not reach the same conclusion. Hence it might prove difficult for theories to become accepted following repeated trials by different scientists. Finally, Bayesians often use "noninformative" priors in order to "let the data speak for themselves," yet there is no unanimity among Bayesians regarding the proper or "correct" noninformative prior distributions to use with common likelihoods, or even on the exact definitions of noninformative and/or vague priors.

Although the concerns detailed in the preceding paragraph are serious, we do not find them compelling and, when balanced against the objections to non-Bayesian methods, we find that Bayesian methods are often preferable. We agree with the oft-stated Bayesian position that, despite claims to the contrary, all scientific work is subjective; just the act of choosing a likelihood function is subjective. To us, Bayesian analysis is more honest because it forces scientists to confront their subjectivity at the outset and not "sweep it under the rug." We believe the possibility that different scientists may reach differing conclusions based on the same data if they bring vastly different prior opinions to the problem is not a disadvantage but rather a description of the true state of the world. Furthermore, given that some basic care is taken not to rule out certain outcomes a-priori, then as evidence accumulates even scientists with markedly differing initial priors will eventually reach the same conclusion (Box and Tiao 1972). A Bayesian analysis is not complete unless the scientist specifies the prior distribution that was used. If a reader does not accept the prior distribution, then they are under no obligation to accept the results, much the same as in a non-Bayesian study, if the reader does not accept the choice of likelihood, they will not find the evidence convincing. Finally, although it is true that there is no universal agreement regarding non-informative prior distributions, a wise and/or pragmatic scientist would repeat an analysis using several noninformative priors. If the analyses agree, then this suggests that the choice of prior is unimportant. If they do not, then the scientist has learned something and should perhaps think more deeply about the parameter in question. We find this to be a good feature.

In our view there are important advantages to Bayesian analysis. First and foremost, it makes sense. As detailed earlier in this chapter, Bayesians make inferences about what they don't know, based on what they observe during an experiment and what they believed prior to observing data. To us, this is a perfect and natural analog to the process of learning by experience. The Bayesian inference system is self-contained and all inferences stem from the calculus of probability, e.g., see Box and Tiao (1972). Unlike common statistical techniques such as acceptance/rejection of hypotheses based on p-values or calculation of confidence intervals, Bayesian inference does not require scientists to imagine an infinite series of repeated trials of an experiment under identical situations. A Bayesian analysis is conditioned on the data you observed, not on data you *might* have observed, e.g., see Lindley (1990) or Carlin and Louis (2009). In a classic series of papers, Berger and colleagues (Berger 1985; Berger and Delampady 1987; Berger and Selke 1987) have shown that the common non-Bayesian practice of rejecting a hypothesis when a p-value takes on a value smaller than some standard value (typically 0.05) may lead to rejection of hypotheses when the posterior probability of the hypothesis being true, given the available evidence, is actually greater than 0.5, i.e., the hypothesis is more likely to be true than not; this is remarkable! In another classic paper, Lindley (1957) showed that a classical hypothesis testing procedure can *reject* a null hypothesis at the $(\alpha)100\%$ level while, incredibly, a Bayesian procedure might conclude that the posterior probability that the null is *true* is $(1-\alpha)100\%$! This has come to be known as Lindley's Paradox. In our experience, the common Bayesian assertion that most scientists do not fully understand p-values is correct. We believe it is true that many scientists mistakenly regard a p-value as the probability that the null hypothesis is true[3], whereas even a cursory look at the development of p-values shows that it is nothing of the sort. Rather it is the probability of observing data as extreme *or more extreme* as the data actually observed *if* the null hypothesis is true. The latter can be vastly different than the former. We agree with Cohen (1994), p. 997 who said of a p-value "... it does not tell us what we want to know, and we so much want to know what we want to know that, out of desperation, we believe that it does!" We find ourselves in accord with Jeffreys (1980) p. 453: "I have always considered the arguments for the use of P absurd. They amount to saying that a hypothesis that may or may not be true is rejected because a greater departure from the trial value was improbable; that is, that it has not predicted something that has not happened." It is worthwhile to note that in March 2016, the American Statistical Association issued a policy statement on p-values which, among other things, argues against their routine use in science (Wasserstein and Lazar 2016).[4]

Finally, in practical situations, non-Bayesian methods may require scientists to violate their own assumptions in order to proceed. In an oft-cited example, Lindley and Phillips (1976) show that in a common, unremarkable experiment involving a series of identical trials, each resulting in one of two outcomes (say heads or tails), a proper non-Bayesian analysis is impossible unless the number of trials in known in advance. If the sample size is not known before the start of the experiment, then it is impossible to clearly define the sample space of possible outcomes; this sample space is required for the computation of p-values. However it is common for studies to be terminated for unanticipated reasons. As a consequence non-Bayesians often violate their own assumptions and proceed *as if* the sample size was known a-priori. This problem does not arise in Bayesian inference.

---

[3]One of the authors actually witnessed a well-respected scientist instruct a candidate in this definition during a graduate committee meeting!

[4]Reliance on p-values is also at least partially responsible for the unsettling lack of reproducibility in scientific studies (e.g., see Begley and Ioannidis 2015). As a partial solution to this, Benjamin et al. (2018) have suggested that p-values between the conventional 0.05 and a much more stringent 0.005 be called *suggestive* and that *significant* findings should be restricted to those with a p-value less than 0.005.

As a result of using a self-contained inference system, Bayesian scientists are relieved of considering ad-hockeries in order to complete analyses. Each analysis follows the same well-defined steps. Furthermore, complicated situations with high-dimensional problems and/or more than one data set are naturally accommodated. Hence rather than considering seemingly arcane mathematics, scientists are freed to focus on choosing the appropriate likelihood and prior distributions—those which most faithfully describe the phenomena under investigation. We believe that this is where the scientist's focus should be.

Given the preceding discussion, it may occur to some scientists to wonder "why wasn't I taught Bayesian statistics as an undergraduate?". While there are a number of possible contributing reasons for this, we believe that a primary impediment to the teaching of Bayes to non-statisticians is the fact that it would require students to first acquire (typically via yet another course) some knowledge of probability and probability distributions. Since science students already must take many courses in their field of study, the prospect of taking a probability course in order to be able to take a Bayesian course is unappealing. Hence students normally opt for one statistical course covering classical methods. This is unfortunate. We regard Bayesian vs. non-Bayesian statistics as roughly analogous to two types of amusement parks: In one there is a significant admission price, but afterwards all the rides are free. This is analogous to Bayesian methods. Before entering, one has to pay the cost of learning probability. Afterwards, the logic of Bayesian methods is empowering. On the other hand, there are amusement parks with no admission cost but a fee for every ride. This is analogous to non-Bayesian methods; it is easy enough to learn simple, basic concepts like t-tests and/or least squares but whenever new situations occur, the scientist must learn a new set of statistical methods.

References

Begley, C. G., & Ioannidis, J. P. A. (2015). Reproducibility in science: Improving the standard for basic and preclinical research. *Circulation Research*, *116*, 116–126.

Benjamin, D. J., Berger, J. O., Johannesson, M., Nosek, B., Wagenmaker, E. J., Berk, R. et al. (2018). Redefine statistical significance. *Nature Human Behavior*, *1*, 6–10.

Berger, J. O. & Delampady, M. (1987). Testing precise hypotheses. *Statistical Science*, *2*, 317–352 (with discussion).

Berger, J. O. & Selke, T. (1987). Testing a point null hypothesis: The irreconcilability of significance levels and evidence. *Journal of the American Statistical Association*, *82*, 112–133 (with discussion).

Berger, J. O. (1985). *Statistical Decision Theory and Bayesian Analysis*. New York, NY: Springer.

Bernardo, J. M., & Smith, A. F. M. (1994). *Statistical Decision Theory and Bayesian Analysis*. New York, NY: Wiley.

Berry, D. A. (1997). Teaching elementary Bayesian statistics with real applications in science. *The American Statistician*, *51*(3), 241–246.

Box, G. E. P., & Tiao, G. C. (1972). *Bayesian Inference in Statistical Analysis*. Reading, MA: Addison-Wesley.

Carlin, B. P., & Louis, T. A. (2009). *Bayesian Methods for Data Analysis* (3rd ed.). Boca Raton, FL: Chapman & Hall/CRC.

Carpenter, B., Gelman, A., Hoffman, M., Lee, D., Goodrich, B., Betancourt, M., et al. (2017). Stan: A probabilistic programming language. *Journal of Statistical Software, Articles, 76*(1), 1–32. https://www.jstatsoft.org/v076/i01.

Casella, G., & Berger, R. L. (2001). *Statistical Inference* (2nd ed.). Belmont, CA: Duxbury Press.

Cohen, J. (1994). The earth is round (p<.05). *AmericanPsychologist, 49*, 997–1001.

de Finetti, B. (1974). *Theory of Probability* (Vol. 1). New York, NY: Wiley.

de Valpine, P., Turek, D., Paciorek, C., Anderson-Bergman, C., Temple Lang, D., & Bodik, R. (2017). Programming with models: Writing statistical algorithms for general model structures with NIMBLE. *Journal of Computational and Graphical Statistics, 26*, 403–413.

Fisher, R. A. (1922). On the mathematical foundations of theoretical statistics. *Philosophical Transactions of the Royal Society of London. Series A, 222*, 309–368.

Gelman, A., Carlin, J. B., Stern, H. B., Dunson, D. B., Vehtari, A., & Rubin, D. B. (2013). *Bayesian Data Analysis* (3rd ed.). New York, NY: Chapman & Hall/CRC.

Green, E. J., Strawderman, W. E., & Airola, T. M. (1992). Assessing classification probabilities for thematic maps. *Photogrammetric Engineering and Remote Sensing, 59*, 635–639.

Harris, B. (1966). *Theory of Probability*. Reading, MA: Addison-Wesley.

Jeffreys, H. (1935). Some tests of significance, treated by the theory of probability. *Mathematical Proceedings of the Cambridge Philosophical Society, 31*(2), 203–222.

Jeffreys, H. (1980). Some general points in probability theory. In A. Zellner (Ed.), *Bayesian Analysis in Econometrics and Statistics* (pp. 451–454). Amsterdam: North-Holland.

Joseph, L., Gyorkos, T. W., & Coupal, L. (1995). Bayesian estimation of disease prevalence and the parameters of diagnostic tests in the absence of a gold standard. *American Journal of Epidemiology, 141*, 263–272.

Kolmogorov, A. N. (1933). *Grundbegriffe der Wahrscheinlichkeitrechnung*. Ergebnisse Der Mathematik; translated as Foundations of Probability. New York, NY: Chelsea Publishing Company.

Lindley, D. V. & Smith, A. F. M. (1972). Bayes estimates for the linear model. *Journal of the Royal Statistical Society. Series B (Methodological), 34*, 1–41 (with discussion).

Lindley, D. V. (1990). The 1988 Wald memorial lectures: The present position in Bayesian statistics. *Statistical Science, 5*, 44–89 (with discussion).

Lindley, D. V. (1957). A statistical paradox. *Biometrika, 44*(1/2), 187–192.

Lindley, D. V., & Phillips, L. D. (1976). Inference for a Bernoulli process (a Bayesian view). *American Statistician, 30*(3), 112–119.

Lunn, D. J., Thomas, A., Best, N., & Spiegelhalter, D. (2000). Bayesian modelling framework: Concepts, structure, and extensibility. *Statistics and Computing, 10*, 325–337.

McGrayne, S. B. (2012). *The Theory that Would Not Die: How Bayes' Rule Cracked the Enigma Code, Hunted Down Russian Submarines, & Emerged Triumphant from Two Centuries of Controversy*. New York and London: Yale University Press.

Plummer, M. (2003). JAGS: A program for analysis of Bayesian graphical models using Gibbs sampling. In *Proceedings of the 3rd international workshop on distributed statistical computing* (Vol. 124, p. 125). Vienna: Austria.

Richards, J. A., & Jia, X. (2006). *Remote Sensing Digital Image Analysis: An Introduction*. New York, NY: Springer.

Robert, C. P. (2001). *The Bayesian Choice: From Decision-Theoretic Foundations to Computational Implementation* (2nd ed.). New York, NY: Springer.

Spiegelhalter, D. J., Myles, J. P., Jones, D. R., & Abrams, K. R. (1999). An introduction to Bayesian methods in health technology assessment. *BMJ, 319*(7208), 508–512.

Stigler, S. (1986). *The History of Statistics*. Cambridge, MA: Belknap Press.

Wasserstein, R. L., & Lazar, N. A. (2016). The ASA's statement on p-values: Context, process, and purpose. *The American Statistician, 70*(2), 129–133. http://www.tandfonline.com/doi/abs/10.1080/00031305.2016.1154108.

Chapter 2
Probability Theory and Some Useful Probability Distributions

2.1 Discrete and Continuous Random Variables

Effective Bayesian inference requires familiarity with probability distributions. In fact, as will be seen in subsequent chapters, the most important choices in Bayesian inference usually involve the choices of distributions to represent the state of knowledge before data is collected (prior distribution) and the sampling distribution, also referred to as the data model, for the observable data. Hence in this chapter we will review some commonly used probability distributions. Readers already knowledgeable about probability distributions should skim this chapter to insure that they are comfortable with our notation and terminology.

In the following, we assume a random variable is a randomly varying quantity whose exact value may or may not be known, and may or may not be discoverable through experimentation or observation. We follow the usual practice of denoting random variables by italicized upper case letters and values of random variables by italicized lower case letters. For instance, the height of trees in a loblolly pine plantation might be denoted as H and particular values by h, so, for example, we would denote the probability that the height of a tree is in the interval (h_1, h_2) by $P(h_1 < H < h_2)$.

It is important to distinguish between discrete and continuous random variables. For our purposes, a *discrete* random variable is one which can assume only certain values in a denumerable set. The values may be bounded or not. For example, the number of deer in a certain area is discrete: there can be 41 or 42 deer, but there cannot be 41.1 or 41.8. Note that this random variable is bounded below by 0, but is effectively unbounded from above. On the other hand, suppose we are studying deer-car collisions on a certain road. We may be interested in the number of days in a month that at least one collision is observed. Suppose the month in question is November of a given year. This discrete random variable is bounded from above and below and must assume one of the values $0, 1, 2, \ldots, 29, 30$.

© Springer Nature Switzerland AG 2020
E. J. Green et al., *Introduction to Bayesian Methods in Ecology and Natural Resources*,
https://doi.org/10.1007/978-3-030-60750-0_2

For discrete random variables, we may properly consider the probability of specific values. For instance, in the latter situation we might ask "what is the probability that the number of days on which at least one collision occurs is 5?" Theoretical probability distributions for discrete random variables may be used to compute the probabilities of particular values, and must sum to one, i.e., if the random variable is X, we must have

$$\sum_x P(X = x) = 1 \qquad (2.1.1)$$

where the sum is over all values for X with nonzero probability, i.e., all possible values for X. In addition, we require

$$P(X = x) \geq 0 \qquad (2.1.2)$$

for all values of X. The probability that X falls in any set, say set A, is calculated as

$$P(X \in A) = \sum_{x \in A} P(X = x). \qquad (2.1.3)$$

For discrete random variables we often use the shorthand $p(x)$ to represent $P(X = x)$.

Continuous random variables are those which are not constrained to assume particular values, and may be bounded or not. For example the height of a tree might be 15.7 m or 15.73 m, and is not constrained to assume discrete values. In this example the random variable is bounded from below (0) but not from above, unless we can specify the maximum possible height. On the other hand, suppose we are interested in the percentage of bald eagle nests in a particular region which produce eaglets in a particular year. This is also a continuous random variable, but it is constrained to lie between 0 and 100; hence it is bounded from above and below. For continuous random variables, the probability of any specific value is zero, but we may consider the probability that the random variable lies in an interval. For example, if tree height is again represented by H, $P(H = 15.73) = 0$, but we may consider $P(15.72 < H < 15.74)$. Note also that since $P(H = 15.72) = 0$ and $P(H = 15.74) = 0$, it doesn't matter whether we use $<$ or $\leq$ in the statement of the interval probability.

It may be off-putting at first to learn that the probability of any specific value is zero for a continuous random variable. This is a result of the way we assess probabilities for such variables. We use mathematical curves or functions to represent probability distributions, and the area under the curve of any interval equals the probability of that interval. Specific values are points and since a point has a width of zero, the area under the curve of a point is zero.

Probability calculations for continuous random variables are computed using what are usually called probability densities. The total area under a probability density equals 1. In Fig. 2.1 we display a continuous random variable (X) and its corre-

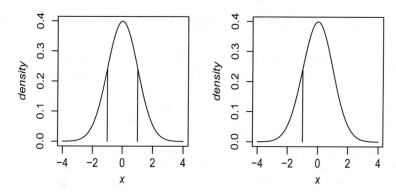

Fig. 2.1 A random variable X and its probability density. In the left graph, the area under the curve between the two vertical lines is equal to the probability that X is between -1 and 1. One the right, we see that the area under the curve corresponding to $X = -1$ is zero

sponding probability density. In the graph on the left we have identified the interval corresponding to $(-1 \le X \le 1)$. The area under the curve in this interval corresponds to $P(-1 \le X \le 1)$. In the graph on the right we have identified the point $X = -1$. Since this point is dimensionless, the area under the curve corresponding to the point is 0 and hence $P(X = -1) = 0$.

We will use $f(\cdot)$ to denote probability density functions for continuous random variables. However, as is common practice, we will use the terms "probability density" and "probability distribution" interchangeably for continuous variables. In context, the meaning will be clear.

As mentioned above, the probability that a continuous variable falls in an interval is equal to the area under the curve of that interval. The latter is found by integrating the curve from the lower limit of the interval to the upper limit of the interval. In the case of Fig. 2.1, $P(-1 \le X \le 1)$ would be computed as

$$P(-1 \le x \le 1) = \int_{-1}^{1} f(x)dx. \qquad (2.1.4)$$

where $f(x)$ is the probability density for X.

We shall say a probability distribution or density has support for a random variable at a particular value (discrete) or range (continuous) if it is non-zero at that value or over that range. For more background on discrete and continuous random variables, and probability distributions, consult an introductory statistics or probability text, e.g., Mendenhall et al. (2008) or Chung and AitSahlia (2003).

2.2 Expectation, Mean, Standard Deviation, and Variance

We will denote the population mean, or expected value, of the random variable X by $E(X)$. If X is a discrete random variable, then the expected value is given by

$$E(X) = \sum_x x \; p(x) \tag{2.2.1}$$

where the sum is over all values of X with nonzero probability. Suppose we toss a fair coin twice. What is the expected number of heads we will observe? Most readers would intuitively guess that the answer is 1. Let's see if Eq. 2.2.1 confirms this... there are four possible outcomes; we can get a head on the first toss and a tail on the second, a tail on the first and a head on the second, a tail on both tosses, or a head on both tosses. Since the coin is fair, we judge each of these outcomes to be equally probable and assign to each a probability of 0.25. Let X be the number of heads we observe. The outcomes, along with their probabilities are shown in Table 2.1. The probability of observing 0 heads is 0.25, the probability of observing just one is 0.5 (because it can come on either the first or second toss), and the probability of observing two is 0.25. So, according to Eq. 2.2.1, the expected number of heads is $(0)(0.25) + (1)(0.5) + (2)(0.25) = 1$, confirming our intuitive answer.

For a continuous random variable, the expected value is given by

$$E(X) = \int_{-\infty}^{\infty} x \; f(x) \; dx \tag{2.2.2}$$

The integral sign in Eq. 2.2.2 is the continuous mathematics analog to the summation sign in Eq. 2.2.1. By integrating from $-\infty$ to ∞ we are sure to include all values of X with support under $f(x)$.

The mean, or expected value, is a common measure of central tendency. Other common measures are the median (middle value in an ordered set, or the 50th percentile) and the mode (most common value). While measures of central tendency tell us something about the typical value of a random variable, they do not tell us anything about how much variation or spread there is in the values. A seemingly obvious way to measure the variation in a random variable might be the average difference between each observation and the mean,

Table 2.1 Possible outcomes from coin tossing experiment

1st toss	2nd toss	Probability	X
T	T	0.25	0
H	T	0.25	1
T	H	0.25	1
H	H	0.25	2

$$E\big(X - E(X)\big) = E\big(X - \mu\big), \tag{2.2.3}$$

if we follow the usual practice of denoting $E(X)$ by μ. Unfortunately, as can easily be shown, this quantity equals zero for all variables and probability distributions. Hence we must seek another expression. One reason why the above expression always equals zero is that positive and negative deviations cancel each other. This may be eliminated by squaring the deviations. The result is the population variance:

$$Var(X) = E\big(X - E(X)\big)^2 = E\big(X - \mu\big)^2. \tag{2.2.4}$$

The variance of a random variable is the average squared deviation about the mean and is commonly denoted by σ^2.

Since the variance is an average squared difference, it is in squared units. For example, if the variable under study is the number of deer in a particular area, the variance is in units of deer^2. To recover the original units of measurement, we compute the standard deviation. This is simply the square root of the variance, and is commonly denoted by σ.

Suppose we have two random variables: X and Y. A common measure of the association between X and Y is given by the covariance, defined as:

$$Cov(X, Y) = E((X - E(X))(Y - E(Y))) = E(XY) - E(X)E(Y). \tag{2.2.5}$$

If X and Y are independent, then $E(XY) = E(X)E(Y)$, and hence $Cov(X, Y) = 0$. The covariance of X and Y is often denoted by σ_{XY}. Unlike the variance and standard deviation, the covariance is not constrained to be non-negative. If two variables are negatively correlated, i.e., large values of one tend to be associated with small values of the other and vice-versa, then their covariance would be negative.

2.3 Unconditional, Conditional, Marginal, and Joint Distributions

Probability densities and/or distributions may be characterized as conditional, unconditional, marginal, and/or joint distributions. Most readers have probably been exposed to common probability distributions, such as the binomial distribution or the normal distribution. Interestingly, standard distributions such as these may be categorized as either unconditional or conditional, depending on whether they are viewed in a non-Bayesian or Bayesian context, respectively.

To a non-Bayesian, a probability distribution fully describes the behavior of a random variable, and such a distribution is therefore thought of as an *unconditional* distribution. The distributions are generally indexed by parameters, but the parameters are considered to be fixed constants. On the other hand, as mentioned in Chap. 1, Bayesians believe that if parameters are unknown, it is legitimate to attach probability distributions to them. In the latter case, the standard, unconditional distributions of the non-Bayesian are interpreted as *conditional* distributions; they describe the behavior of the random variable conditional on the value(s) of the parameter(s).

The objective of a Bayesian analysis is usually to estimate the posterior distribution of parameters of interest, given the data and what was known about the parameters before any data were observed. Suppose the random variable X follows a density indexed by two parameters: μ and σ^2. Then the objective of a Bayesian analysis might be to estimate $f(\mu, \sigma^2 \mid x)$, i.e., the probability distribution of μ and σ^2 conditional on the observed data, x. This is called a *joint conditional* distribution because it jointly describes the behavior of both μ and σ^2, given x. If we were only interested in one of the parameters, say μ, then we might operate mathematically on the joint conditional distribution to remove σ^2 and obtain the *marginal conditional* distribution of μ given x, $f(\mu \mid x)$. Readers should consult any elementary statistics or probability text for more details on marginal and joint distributions.

2.4 Likelihood Functions and Random Samples

As described in Chap. 1, the likelihood function plays an important role in Bayesian analysis. Bayes theorem states $P(\theta \mid x) \propto L(\theta \mid x) P(\theta)$, and hence it is through the likelihood function L that the observed data x influence the posterior distribution. We will use the terms *likelihood function, sampling distribution*, and *data model* interchangeably. Readers have almost certainly heard or read that statisticians prefer data to arise from a random sample. A random sample of size n is one in which every possible sample of size n has an equal and independent probability of being observed. An immediate consequence of this is that the individual observations in a sample are independent. The primary reason for preferring that data arise from random samples is that it makes the job of computing the likelihood function relatively easy, whereas for a non-random sample, it may be much more difficult.

Recall from Chap. 1 that the likelihood function is of the same form as $P(x \mid \theta)$, except that in the case of the likelihood function, we regard the expression as conditional on the data x instead of on the parameter θ. It bears repeating that this is done in both Bayesian and non-Bayesian statistical methods. Hence given a random sample, the task is to formulate $P(x \mid \theta)$. Thus far we have used a generic x to denote the observed data. Now we'll get more specific: suppose we observe a random sample of size n: $x_1, x_2, \ldots, x_n$. A short-hand way of referring to the n values of X is as a vector, denoted by x, i.e., $x = (x_1, x_2, \ldots, x_n)^\top$, where $\top$ is the transpose operator. We can invoke the multiplicative rule of probability (1.2.1), and write

$$P(x \mid \theta) = P(x_n \mid x_{n-1}, x_{n-2}, \ldots, \theta) \times P(x_{n-1} \mid x_{n-2}, x_{n-3}, \ldots, \theta) \times \ldots$$
$$\times P(x_2 \mid x_1, \theta) \times P(x_1 \mid \theta). \quad (2.4.1)$$

Since the sample was specified to be a random sample, the observations are independent and (2.4.1) reduces to the much more manageable

$$P(x \mid \theta) = P(x_n \mid \theta) \times P(x_{n-1} \mid \theta) \times \ldots \times P(x_2 \mid \theta) \times P(x_1 \mid \theta) = \prod_{i=1}^{n} P(x_i \mid \theta).$$

$$(2.4.2)$$

The mathematical operator on the RHS of (2.4.2) is the product operator. It is the multiplicative analog of the usual summation operator, Σ. Hence, if the probability distribution of the sample observations is known, the likelihood function for a random sample of size n is:

$$L(\theta \mid x) = \prod_{i=1}^{n} P(x_i \mid \theta). \qquad (2.4.3)$$

We will use $p(x \mid \theta)$ to denote probability distributions for discrete random variables and $f(x \mid \theta)$ to denote probability densities for continuous random variables.

2.5 Some Useful Discrete Probability Distributions

2.5.1 Binomial Distribution

Suppose a random variable Y can assume only one of two values, for example (yes, no), (alive, dead) or (male, female). For lack of better terminology, call one value of Y a success and the other a failure. Now suppose Y is observed a number of times. Each observation will be called a trial. Trials with only two permissible outcomes, and for which the probability of success remains constant from trial to trial, are often called *Bernoulli* trials, after the 17th Century Swiss mathematician Jacob Bernoulli. We shall call a series of such trials an experiment. Let X be the total number of successes observed in the experiment. Then X follows a binomial distribution if the following three conditions hold:

1. The probability of success is constant from trial to trial;
2. The trials are independent;
3. The number of trials is determined before the experiment is begun.

The probability distribution for X is conditioned on two parameters: n and p, where n is the number of trials and p is the probability of success on a single trial, and is defined mathematically as:

$$p(x \mid n, p) = \binom{n}{x} p^x (1-p)^{n-x} = \frac{n!}{x!(n-x)!} p^x (1-p)^{n-x} \qquad (2.5.1)$$

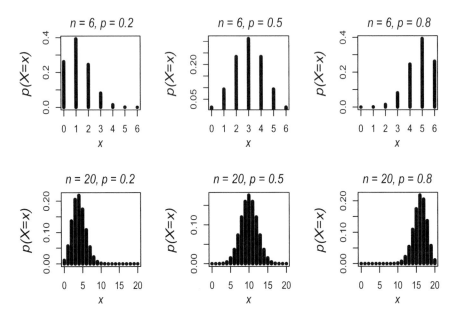

Fig. 2.2 The binomial probability distribution for selected values of n and p.

where, for any integer m, $m!$ is called m-factorial, defined as $m! = (m)(m-1)(m-2)\ldots(1)$, and by definition, $0! = 1$. The mean of X is np, and the variance is $np(1-p)$ or npq, where $q = 1 - p$. Note that while X is constrained to take on integer values, its mean and variance are not. We use the notation $X \sim \mathbf{Bi}(n, p)$ to indicate that X follows the binomial distribution with parameters n and p.

Figure 2.2 displays the binomial distribution for a few values of n and p. When p is small (low probability of success), the distribution is skewed to the right; the total number of successes is usually low. The reverse happens when p is large. When p is 0.5, successes and failures are equally likely and the distribution is symmetrical.

The binomial distribution is widely used when the phenomenon under study falls into one of two classes; e.g., is a plant healthy or not; is a stream polluted or not; is the subject alive or dead; male or female; fertile or infertile, etc. The assumptions of the binomial distribution are usually not *exactly* met, yet the binomial distribution often provides an adequate description of the data. In a Bayesian context, the binomial distribution is usually used as a likelihood, or sampling distribution.

2.5.2 Multinomial Distribution

Suppose, as in the previous section, we are studying a discrete random variable, but now we permit the number of outcomes to be greater than two. For example, insect damage to a tree may be none, light, moderate, or severe. Let the number of possible outcomes be denoted by k. Suppose the probability that an independently drawn observation assumes value j is given by p_j, $j = 1, 2, \ldots, k$. Note we must have $\Sigma_{j=1}^{k} p_j = 1$. Once again we require that the observations are independent, the probabilities remain constant from trial to trial, and that the total number of trials (n) is known in advance. This is called a multinomial experiment, and the probability of observing x_1 trials resulting in outcome 1, x_2 trials resulting in outcome 2, $\ldots$, x_k trials resulting in outcome k is given by

$$p(x_1, x_2, \ldots, x_k \mid n, \boldsymbol{p}) = p(\boldsymbol{x} \mid n, \boldsymbol{p}) = \binom{n}{x_1, x_2, \ldots, x_k} p_1^{x_1} p_2^{x_2} \cdots p_k^{x_k}$$

$$= \frac{n!}{x_1! x_2! \ldots x_k!} p_1^{x_1} p_2^{x_2} \cdots p_k^{x_k}. \quad (2.5.2)$$

Here, $\boldsymbol{x}$ and $\boldsymbol{p}$ represent the vectors (or columns) of observations and probabilities, respectively. In this case the length of the vector is k, i.e., each potential outcome j, $j = 1, 2, \ldots, k$, has a probability. Of course, we must have $0 \leq x_j \leq n$, $j = 1, 2, \ldots, k$, and $\Sigma_{j=1}^{k} x_j = n$.

The expected number of outcomes in category j is given by np_j with a variance of $np_j(1 - p_j)$. It is worth noting that the distribution of X_j is $\mathbf{Bi}(n, p_j)$ for each j, $j = 1, 2, \ldots, k$. We use the notation $X \sim \mathbf{Mu}(n, \boldsymbol{p})$ to indicate that the vector X follows the multinomial distribution with parameters n and $\boldsymbol{p}$. The multinomial distribution is used extensively in capture-recapture studies, where the classes are the various capture histories an animal might exhibit (Williams et al. 2002). Another use of the multinomial distribution is land-use classification (Green et al. 1992). In a Bayesian context, the multinomial distribution is usually used as a likelihood, or sampling distribution.

2.5.3 Poisson Distribution

The Poisson distribution is commonly used for count data[1], such as the number of vehicle accidents at a particular road intersection in a given time interval, or the number of plants of a certain species on a plot of specified size. Observations from the Poisson distribution are bounded below by zero, but are unbounded from above.

If the random variable X follows the Poisson distribution, then the probability that $X = x$ is given by

$$p(x \mid \lambda) = \frac{\lambda^x e^{-\lambda}}{x!}, \quad x = 0, 1, 2, \ldots; \lambda > 0. \quad (2.5.3)$$

---

[1] We posit that the Poisson is the *standard* distribution for count data and ought to be used unless there is a compelling reason not to.

The Poisson distribution is indexed by only one parameter: λ. This is sometimes referred to as the rate parameter (especially if the random variable under study is related to counts during a specified time interval or a similar phenomenon). We use the notation $X \sim \mathbf{Poi}(\lambda)$ to indicate that X follows the Poisson distribution with parameter λ.

The Poisson distribution possesses at least two useful properties. First, the mean and variance of X are identical; both are equal to λ. Second, if X follows a Poisson distribution with parameter λ_X and Y follows a Poisson distribution with parameter λ_Y and X and Y are independent, then $(X + Y)$ follows a Poisson distribution with parameter $(\lambda_X + \lambda_Y)$.

Figure 2.3 displays the Poisson distribution for a few values of λ. Note that the distribution becomes more symmetrical as λ increases. In a Bayesian context, the Poisson distribution is usually used as a likelihood, or sampling distribution.

The Poisson distribution can be derived from the binomial distribution. Recall that the mean of the Poisson distribution is λ and the mean of the binomial distribution is np. It can be shown that if we set $\lambda = np$, then substitute $\frac{\lambda}{n}$ for p in the binomial distribution (Eq. 2.5.1) and take the limit as n approaches ∞, the result is the Poisson distribution (Eq. 2.5.3). But note that in doing so, we have $p = \frac{\lambda}{n}$. So as n goes to ∞, p gets very small. This has unfortunately led to the Poisson being termed a "distribution of rare events." In point of fact, a variable modeled by a Poisson distribution need not be rare at all. For example, suppose the number of plants per mil-acre plot of a certain species follows a Poisson distribution with $\lambda = 3$. This

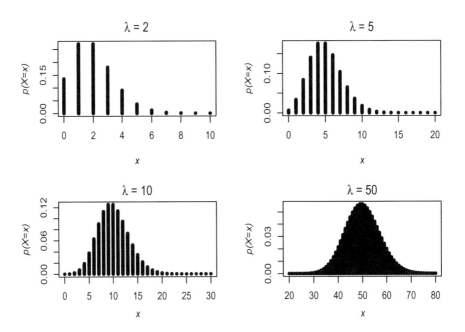

Fig. 2.3 The Poisson probability distribution for selected values of λ

is a perfectly reasonable scenario, and indicates that the average number of plants per plot is 3. Hence these plants are not "rare" in any common sense of the term. In fact, when there are too many 0's in the data (which one would expect with a "rare" phenomenon), the Poisson distribution is often not a good fit, and might be augmented to a "zero-inflated Poisson" (ZIP) model (e.g., see Lambert 1992).

2.6 Some Useful Continuous Probability Distributions

2.6.1 Uniform Distribution

The uniform distribution is the simplest continuous probability distribution. Suppose the random variable X is bounded from below by a and from above by b. Further suppose that all intervals between a and b of the same length are equally likely. Then X follows the uniform distribution with parameters a and b. The uniform density is defined as

$$f(x \mid a, b) = \frac{1}{b - a}, \quad a \leq x \leq b, \tag{2.6.1}$$

The mean and variance of the uniform distribution are given by $(a + b)/2$ and $(b - a)^2/12$, respectively. We use the notation $X \sim \mathbf{Unif}(a, b)$ to indicate that X follows the uniform distribution with parameters a and b. The standard uniform distribution is a special case of the uniform distribution with $a = 0$ and $b = 1$. This is the basic distribution used in random number generation. The uniform distribution is shown in Fig. 2.4, with $a = 2$ and $b = 4$. This distribution is also sometimes referred to as a flat or rectangular distribution, and the reason for these names is clear from Fig. 2.4. In Bayesian analysis, the uniform distribution is often used as a prior distribution to represent ignorance about potential values of an unknown parameter (of course, we must have *some* knowledge about potential values in order to specify a and b; hence the uniform distribution is usually considered to represent vague knowledge rather than no knowledge at all).

2.6.2 Normal Distribution

The normal distribution, also known as the Gaussian distribution, is arguably the most familiar continuous probability distribution. This distribution is indexed by two parameters, the mean (μ) and variance (σ^2). We use the notation $X \sim \mathbf{N}(\mu, \sigma^2)$ to indicate that X follows the normal distribution with parameters μ and σ^2. The mean is called the location parameter because it defines the position of the center of the distribution on the real number line, and the variance is called the scale parameter

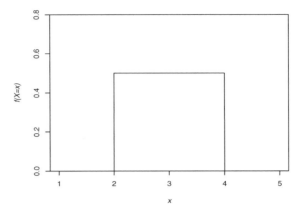

Fig. 2.4 The Uniform probability density for $a = 2$ and $b = 4$

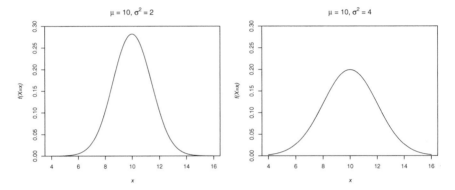

Fig. 2.5 The Normal probability density for $\mu = 10, \sigma^2 = 2$ and $\mu = 10, \sigma^2 = 4$

because it defines how "stretched-out" the distribution is. Given these two parameters, the distribution takes the form of the familiar "bell curve" as shown in Fig. 2.5 for $\mu = 10, \sigma^2 = 2$ and $\mu = 10, \sigma^2 = 4$. The influence of the scale parameter is easily seen; the distribution with the larger variance has wider tails and a lower value at the mean; hence it has a stretched-out appearance. It is worth noting that although the normal distribution is often thought of as synonymous with the term bell curve, it is not the only distribution to assume a bell-shaped form. Many distributions either do or can display a unimodal, symmetric bell-shaped form.

The normal distribution is defined as

$$f(x \mid \mu, \sigma^2) = \frac{1}{\sigma\sqrt{2\pi}} e^{-(x-\mu)^2/2\sigma^2}, \; -\infty < x < \infty; \; \sigma > 0; \; -\infty < \mu < \infty.$$

$$(2.6.2)$$

It is worthwhile to review a few characteristics of the normal distribution:

- The distribution is defined for all values of x and μ.
- The variance (σ^2) must be positive.
- The distribution is symmetric about μ, and μ is the mean, median and mode.
- When $\mu = 0$ and $\sigma^2 = 1$, the distribution is called the standard normal distribution, and the random variable is often denoted by Z with values represented as z.

The normal distribution is widely used in Bayesian and non-Bayesian statistical inference. However, its widespread use does not derive from a belief that many or most random variables are normally distributed (although at one time this was thought to be true; e.g., see Salsburg 2001). Rather its use is primarily due to the Central Limit Theorem which roughly states that regardless of the underlying distribution of X, as the size n of a random sample increases, the distribution of the sample mean approaches a normal distribution with mean μ and variance σ^2/n, where μ and σ^2 are the mean and variance of X, respectively (e.g., see Bickel and Doksum 2000). This relieves scientists from specifying the distribution of X in the presence of adequate sample sizes and permits the use of a rich body of theory based on the normal distribution. In a Bayesian context, the normal distribution is commonly used as both a likelihood and prior distribution.

2.6.3 Multivariate Normal Distribution

In general, let $v = (v_1, v_2, \ldots, v_k)^\top$ denote the transpose of a k-dimensional column vector, i.e.,

$$v = \begin{bmatrix} v_1 \\ v_2 \\ \vdots \\ v_k \end{bmatrix}, \tag{2.6.3}$$

and let $\mathbf{A}_{k \times k}$ denote a $k \times k$ positive definite matrix, i.e.,

$$\mathbf{A}_{k \times k} = \begin{bmatrix} a_{11} & a_{12} & \ldots & a_{1k} \\ a_{21} & a_{22} & \ldots & a_{2k} \\ \vdots & \vdots & & \vdots \\ a_{k1} & a_{k2} & \ldots & a_{kk} \end{bmatrix}, \tag{2.6.4}$$

where a_{ij} is the value in row i, column j. Furthermore, let det$\mathbf{A}$ indicate the *determinant* of $\mathbf{A}$, and let $\mathbf{A}^{-1}$ denote the *inverse* of $\mathbf{A}$ (for more details on the determinant, inverse, and matrix operations consult a standard linear algebra text such as Bretscher 2004). Suppose we have k random variables $(X_1, X_2, \ldots, X_k)$, and the vector $X = (X_1, X_2, \ldots, X_k)^\top$ follows the joint distribution

$$f(x \mid \boldsymbol{\mu}, \boldsymbol{\Sigma}) = \frac{1}{(2\pi)^{k/2}(\det\boldsymbol{\Sigma})^{1/2}} e^{-\frac{1}{2}(x-\mu)^{\top}\boldsymbol{\Sigma}^{-1}(x-\mu)}, \tag{2.6.5}$$

where $\boldsymbol{\Sigma}$ is positive definite and the density is defined for $-\infty < x_i, \mu_i < \infty$, $i = 1, 2, \ldots, k$. The vector $\boldsymbol{\mu}$ is called the mean vector and $\boldsymbol{\Sigma}$ is called the covariance matrix and the density in (2.6.5) is called the multivariate normal distribution, denoted as $X \sim \mathbf{MVN}(\boldsymbol{\mu}, \boldsymbol{\Sigma})$. If the X_i's are independent then all the off-diagonal terms in $\boldsymbol{\Sigma}$ are equal to 0, i.e.,

$$\boldsymbol{\Sigma} = \begin{bmatrix} \sigma_1^2 & 0 & \cdots & 0 \\ 0 & \sigma_2^2 & \cdots & 0 \\ \vdots & \vdots & & \vdots \\ 0 & 0 & \cdots & \sigma_k^2 \end{bmatrix}. \tag{2.6.6}$$

If the X_i's are *not* independent then the off-diagonal terms represent the covariances, i.e.,

$$\boldsymbol{\Sigma} = \begin{bmatrix} \sigma_1^2 & \sigma_{12} & \cdots & \sigma_{1k} \\ \sigma_{12} & \sigma_2^2 & \cdots & \sigma_{2k} \\ \vdots & \vdots & & \vdots \\ \sigma_{1k} & \sigma_{2k} & \cdots & \sigma_k^2 \end{bmatrix}. \tag{2.6.7}$$

Note $\sigma_{ij} = \sigma_{ji}$.

A plot of the multivariate normal distribution for $k = 2$, with $X_1 = X$, $X_2 = Y$, $\boldsymbol{\mu} = (0, 0)^{\top}$, $\sigma_X^2 = 2$, $\sigma_Y^2 = 1$, and $\sigma_{XY} = 0.5$ is displayed in Fig. 2.6. As with the univariate normal distribution, the multivariate normal distribution is commonly used as both a likelihood and prior distribution in Bayesian analyses. Under the multivariate normal distribution, each of the k individual random variables follows a univariate normal distribution.

2.6.4 t Distribution

The t distribution is a symmetric bell-shaped curve similar to the normal distribution. This distribution was first published by Gosset in 1908 under the pseudonym "Student" (Student 1908), hence it is sometimes referred to as Student's t distribution. In its usual form, the t distribution is centered at 0, i.e., the mean, median, and mode are all equal to 0. In this form, the t distribution is indexed by one parameter, ν, usually called the degrees of freedom. The variance of the t distribution is equal to $\nu/(\nu - 2)$. The t distribution is defined as

Fig. 2.6 Multivariate
normal density for (X, Y)
with $\mu = (0, 0)$, $\sigma_X^2 =$
1, $\sigma_Y^2 = 2$, and $\sigma_{X,Y} = 0.5$

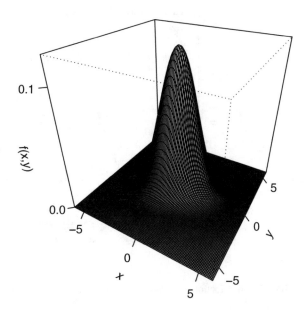

$$f(x \mid \nu) = \frac{\Gamma((\nu + 1)/2)}{\sqrt{\nu \pi} \; \Gamma(\nu/2)} \left(1 + \frac{x^2}{\nu}\right)^{-\left(\frac{\nu+1}{2}\right)}, \quad -\infty < x < \infty, \; \nu > 0, \qquad (2.6.8)$$

where $\Gamma(\cdot)$ is the gamma function (e.g., see Feller 1966). As ν increases, the t distribution converges to the standard normal distribution. The t distribution with $\nu = 5$ is overlaid on the standard normal distribution in Fig. 2.7. Note how similar the two distributions are. When $\nu > 30$, the two distributions are virtually indistinguishable.

The t distribution can be derived as the ratio of X_1 to $\sqrt{X_2/\nu}$ where X_1 follows the standard normal distribution, X_2 follows a chi-squared distribution with ν degrees of freedom and X_1 and X_2 are independent. This fact is used to great effect in non-Bayesian statistics to test hypotheses about sample means (e.g., see Wackerly et al. 2008).

A more general version of the t distribution has three parameters, and is defined as

$$f(x \mid \mu, \nu, \lambda) = \frac{\Gamma((\nu + 1)/2))}{(\nu \pi)^{1/2} \Gamma(\nu/2) \sigma} \left(1 + \frac{1}{\nu}\left(\frac{(x - \mu)}{\sigma}\right)^2\right)^{-\left(\frac{\nu+1}{2}\right)}, \qquad (2.6.9)$$

$$-\infty < x < \infty, \; \nu > 0, -\infty < \mu < \infty, \; \sigma > 0.$$

In Eq. 2.6.9, the distribution is centered at μ and has a mean of μ for $\nu > 1$ and a variance of $(\nu/(\nu - 2))\sigma^2$ for $\nu > 2$. We use the notation $X \sim t_\nu$ and $X \sim t_\nu(\mu, \sigma^2)$ to indicate that X follows the distributions given in (2.6.8) and (2.6.9), respectively.

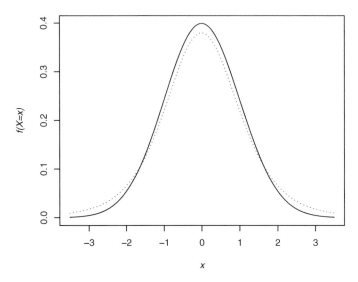

Fig. 2.7 t distribution with $\nu = 5$ (dotted line) and standard normal distribution (solid line)

2.6.5   Gamma Distribution

The gamma distribution is a flexible, two-parameter distribution, with support for positive values. Mathematically, it is defined as

$$f(x \mid \alpha, \beta) = \frac{x^{\alpha-1}\beta^{\alpha}e^{-x\beta}}{\Gamma(\alpha)}, \quad x > 0; \ \alpha > 0; \ \beta > 0. \tag{2.6.10}$$

We use the notation $X \sim \mathbf{Ga}(\alpha, \beta)$ to indicate that X follows the gamma distribution with parameters α and β. Both parameters are constrained to be positive. The first parameter, α, is usually called the shape parameter, while the second parameter, β, is called the rate parameter; β^{-1} is often called the scale parameter[2]. The mean of the gamma distribution is (α/β) and the variance is (α/β^2). The gamma distribution is displayed for some selected parameter values in Fig. 2.8. This distribution has several useful properties; if $X_1 \sim \mathbf{Ga}(\alpha_1, \beta)$, $X_2 \sim \mathbf{Ga}(\alpha_2, \beta)$, and $Y = X_1 + X_2$, then $Y \sim \mathbf{Ga}(\alpha_1 + \alpha_2, \beta)$. Also, if $X \sim \mathbf{Ga}(\alpha, \beta)$ and k is a positive constant, then $kX \sim \mathbf{Ga}(\alpha, k\beta)$. There are several special cases of the gamma distribution, and we present two here:

1. If $X \sim \mathbf{Ga}(1, \beta)$, then X follows the exponential distribution with parameter β, i.e., $X \sim \mathbf{exp}(\beta)$.

---

[2]We define the gamma distribution in terms of the shape and rate parameters. Some authors define it in terms of the shape and scale parameters. Readers should always use caution when using the gamma distribution for inference.

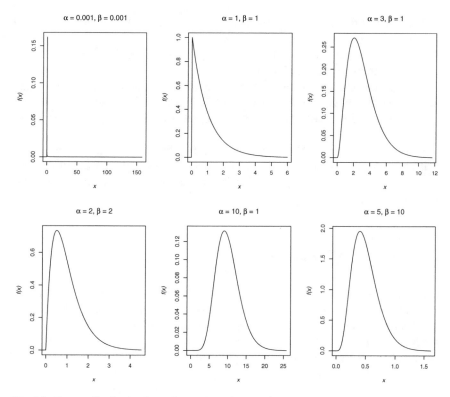

Fig. 2.8 Gamma distribution for various values of α and β

2. If $X \sim \mathbf{Ga}(\nu/2, 1/2)$, then X follows the chi-squared distribution with degrees of freedom $= \nu$, i.e., $X \sim \chi_\nu^2$.

Consult any mathematical statistics text (e.g., Hogg 1978) for more information on the exponential or chi-squared distributions.

The gamma distribution may be a good approximation to the probability density of a random variable that is constrained to be positive and whose distribution is thought to be unimodal and perhaps asymmetric. In a Bayesian context, the gamma distribution is commonly used as both a likelihood and prior distribution. In particular, the gamma distribution is often selected as a prior for a Poisson likelihood.

Sometimes we write $X \sim \mathbf{Ga}^{-1}(\alpha, \beta)$. This means that X follows the *inverse-gamma distribution* with parameters α and β. The density of the inverse gamma distribution is:

$$f(x \mid \alpha, \beta) = \frac{\beta^\alpha}{\Gamma(\alpha) x^{\alpha+1}} e^{-\beta/x}, \quad x > 0; \ \alpha > 0; \ \beta > 0. \tag{2.6.11}$$

The mean of the inverse-gamma distribution is $\beta/(\alpha - 1)$, provided $\alpha > 1$, and the variance is $\beta^2/\left((\alpha - 1)^2(\alpha - 2)\right)$, provided $\alpha > 2$. It is important to note that if X follows the inverse-gamma distribution with parameters α and β, then X^{-1} follows a gamma distribution, also with parameters α and β.

An important special case of the inverse-gamma distribution is the scaled inverse-chi-squared distribution. If $X \sim \mathbf{Ga}^{-1}(\nu/2, (\nu\tau^2)/2)$, then we say X follows the scaled inverse-chi-squared distribution with degrees of freedom = ν and scale τ, i.e., $X \sim \chi_\nu^{-2}(\tau^2)$.

In a Bayesian context, the inverse-gamma distribution is often used as a prior distribution for the variance of a variable with a normal sampling distribution. This is equivalent to specifying a gamma distribution for the *precision*, a pseudonym for the inverse of the variance.

2.6.6 Wishart Distribution

The Wishart distribution is a multivariate generalization of the gamma distribution. It is indexed by two parameters: ρ and $\mathbf{R}$. We use the notation $\mathbf{S} \sim \mathbf{W}_k(\rho, \mathbf{R})$ to indicate that the $k \times k$ matrix $\mathbf{S}$ follows the Wishart distribution with parameters ρ and $\mathbf{R}$, where ρ is a scalar and $\mathbf{R}$ is a $k \times k$ matrix. The exact density of the Wishart distribution may be found in advanced probability and/or statistical texts, e.g., Rao (1973). In a Bayesian context, the Wishart distribution is commonly used as a prior distribution for the inverse of the covariance matrix of a variable with a multivariate normal sampling distribution. The prior estimate of the inverse of the covariance matrix is given by $\rho\mathbf{R}$, and the parameter ρ is called the degrees of freedom. The latter parameter is often interpreted as a measure of strength of the a-priori belief that the inverse of the covariance matrix is close to $\rho\mathbf{R}$.

2.6.7 Beta Distribution

The beta distribution is a two-parameter distribution, with support for values constrained to the interval $(0, 1)$. Both parameters are called shape parameters. We use the notation $X \sim \mathbf{Be}(\alpha, \beta)$ to indicate that X follows the beta distribution with parameters α and β. Mathematically, the beta distribution is defined as

$$f(x \mid \alpha, \beta) = \frac{\Gamma(\beta + \alpha)}{\Gamma(\alpha) + \Gamma(\beta)} x^{\alpha-1}(1 - x)^{\beta-1}, \quad 0 < x < 1; \ \alpha > 0; \ \beta > 0.$$

$$(2.6.12)$$

The mean of the beta distribution is $\alpha/(\alpha + \beta)$, and the variance is $(\alpha\beta)/\left((\alpha + \beta)^2(\alpha + \beta + 1)\right)$. The beta distribution is displayed in Fig. 2.9 for selected values of

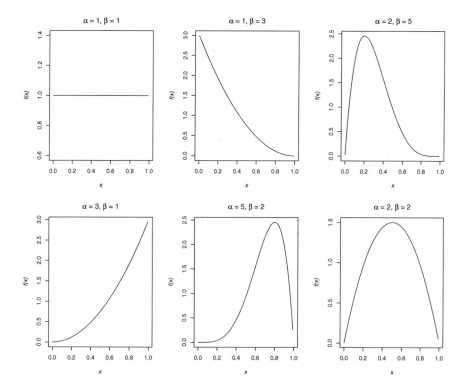

Fig. 2.9 Beta distribution for various values of α and β

α and β. As can be seen in Fig. 2.9, the beta distribution is flexible; it can be symmetric or asymmetric, and can be skewed to the right or left. This distribution is useful for proportions and other quantities constrained to be between 0 and 1. In a Bayesian analysis, the beta distribution is commonly used a prior distribution for parameters constrained to lie in the interval (0, 1) such as p in the binomial distribution.

2.6.8   Dirichlet Distribution

The Dirichlet distribution is a multivariate generalization of the beta distribution. It is useful if we have k continuous random variables, where each is constrained to be in the interval (0, 1), and the k values must sum to 1. Mathematically, the Dirichlet distribution is defined as

$$f(x_1, x_2, \ldots, x_k \mid \alpha_1, \alpha_2, \ldots, \alpha_k) = \frac{\Gamma\left(\sum_{i=1}^{k} \alpha_i\right)}{\prod_{i=1}^{k} \Gamma(\alpha_i)} \prod_{i=1}^{k} x_i^{\alpha_i - 1},$$

$$x_i \geq 0; \quad \sum_{i=1}^{k} x_i = 1; \quad \alpha_i \geq 0. \qquad (2.6.13)$$

Let $\alpha_0 = \sum_{i=1}^{k} \alpha_i$. Then the mean of X_i is (α_i/α_0) and the variance is $\alpha_i(\alpha_0 - \alpha_i)/(\alpha_0^2(\alpha_0 + 1))$. The covariance of X_i and X_j is given by $-\alpha_i\alpha_j/(\alpha_0^2(\alpha_0 + 1))$. We use the notation $X \sim \mathbf{Dir}(\alpha_1, \alpha_2, \ldots, \alpha_k)$ to indicate that $X = (X_1, X_2, \ldots, X_k)^\top$ follows a Dirichlet distribution with parameters $\alpha_1, \alpha_2, \ldots, \alpha_k$. In a Bayesian analysis, the Dirichlet is often used as a prior distribution for $p_i, i = 1, 2, \ldots, k$, where the p_i's are the probabilities from a multinomial distribution. When used as such, it is often helpful to regard α_0 as a "prior sample size" and let this parameter govern the strength of the belief in the prior specifications of $p_i, i = 1, 2, \ldots, k$.

2.7 Exercises

1. Suppose that probability of a randomly chosen individual being a smoker is 0.2. What is the probability that among 2 randomly chosen people we observe 0 smokers, 1 smoker and 2 smokers. Specify the sample space (all possible outcomes) and determine the probability of each outcome. Do the same for three randomly chosen individuals, and find the probability of observing 0, 1, 2, and 3 smokers.
2. Find the probabilities of each outcome in the question 1 by using the binomial distribution.
3. If Y is a random variable with a Poisson distribution satisfying $P(Y = 0) = P(Y = 1)$, what is the mean of Y?
4. Adapt the R script `plot 3 normal densities` (available online) to plot the density for normal distributions with

    (a) $\mu = 10, \sigma^2 = 5$
    (b) $\mu = 20, \sigma^2 = 5$
    (c) $\mu = 10, \sigma^2 = 20$

    Describe the differences between the densities in (a) and (b), and between those in (a) and (c).
5. Suppose Y is binomially distributed with parameters n and p; further suppose the mean of Y is 5 and the variance is 4. Find n and p.
6. Suppose the number of plants per plot follows a Poisson distribution. We randomly install 5 plots and observe the following counts per plot: 12, 8, 9, 9, 12. Your task is to determine the maximum likelihood estimate for the Poisson parameter λ. Normally this would be done using calculus, but here we will do it the brute force way: Calculate $P(y_1, y_2, y_3, y_4, y_5 \mid \lambda)$ for the following values of λ: 8, 9, 9.5, 10, 10.5, 11, 12. For which value of λ is $P(y_1, y_2, y_3, y_4, y_5 \mid \lambda)$ maximized?

References

Bickel, P. J. & Doksum, K. A. (2000). *Mathematical Statistics: Basic ideas and Selected Topics* (Vol. 1, 2nd Edn.). Englewood Cliffs, NJ: Prentice-Hall.

Bretscher, O. (2004). *Linear Algebra with Applications* (3rd Edn.). Englewood Cliffs, NJ: Prentice-Hall.

Chung, K. L., & AitSahlia, F. (2003). *Elementary Probability Theory with Stochastic Processes and an Introduction to Mathematical Finance* (4th Edn.). New York, NY: Springer.

Feller, W. (1966). *An Introduction to Probability Theory and Its Application* (Vol. 2). New York, NY: Wiley & Sons.

Green, E. J., Strawderman, W. E., & Airola, T. M. (1992). Assessing classification probabilities for thematic maps. *Photogrammetric Engineering and Remote Sensing*, *59*, 635–639.

Hogg, R. V., & Craig, A. T. (1978). *Introduction to Mathematical Statistics* (4th Edn.). New York, NY: Macmillan.

Lambert, D. (1992). Zero-inflated Poisson regression, with an application to defects in manufacturing. *Technometrics*, *34*(1), 1–14.

Rao, C. R. (1973). *Linear Statistical Inference and Its Applications* (2nd Edn.). New York, NY: Wiley.

Salsburg, D. (2001). *The Lady Tasting Tea: How Statistics Revolutionized Science in the Twentieth Century*. New York, NY: Henry Holt and Company.

Student. (1908). The probable error of a mean. *Biometrika*, *6*(1), 1–25.

Wackerly, D., Mendenhall, W., & Scheaffer, R. L. (2008). *Mathematical Statistics with Applications* (7th Edn.). New York, NY: Duxbury Press.

Williams, B. K., Nichols, J. D., & Conroy, J. M. (2002). *Analysis and Management of Animal Populations*. San Diego, CA: Academic Press.

Chapter 3
Choice of Prior Distribution

Selecting a prior distribution is integral to Bayesian analyses. In this chapter, we discuss several approaches to specifying priors. First, we discuss the concept of "noninformative" priors. Next we introduce improper priors. Following this, we define conjugate priors. We conclude with a brief discussion of how a scientist might specify an informative prior.

3.1 Vague Prior Distributions

It is often desirable to "let the data speak for themselves" with regard to one or more parameters. For instance, in studies involving the status of resources on public or government-owned lands, it may not be appropriate for a scientist to use thier subjective knowledge. In other situations a scientist may have prior information on one or more parameters, but not on others. Suppose the sample data follow a normal distribution. It may be possible to specify a reasonable prior for the population mean but difficult to specify a prior for the variance. Prior distributions designed to allow the data to dictate inference are variously known as flat, diffuse, vague, noninformative, or uninformative priors. There is some debate in the literature regarding when priors are completely non-informative in the sense of having *no* influence on the posterior distribution (e.g, see Gelman et al. 2013). We are satisfied with Box and Tiao's suggestion that a noninformative prior should be one that "expresses ignorance relative to information which can be supplied by a particular experiment." (Box and Tiao 1972, p. 46). We also prefer the term vague priors to avoid the implication that our priors are completely noninformative.

Berger (2006) (and elsewhere) uses the term "Objective" Bayesian analysis to describe Bayesian analyses which do not entail introduction of subjective information in prior distributions. Among the advantages Berger lists for Objective Bayesian analysis are that highly complex problems can be readily handled with MCMC

© Springer Nature Switzerland AG 2020
E. J. Green et al., *Introduction to Bayesian Methods in Ecology and Natural Resources*,
https://doi.org/10.1007/978-3-030-60750-0_3

Bayesian sampling schemes, that different data sources are naturally accommodated, and that multiple comparisons are easily handled. Berger (2006) advocates the use of "reference priors" for implementing Objective Bayesian analysis. Reference priors were introduced by Bernardo (1979). A key feature of reference priors is that the scientist must define the parameters of interest, and the priors are designed to be noninformative with respect to those. However, reference priors may be improper probability distributions. As described in the next section, this can be an impediment to the routine use of reference priors.

3.2   Improper Prior Distributions

Valid Bayesian inference requires a proper posterior distribution. Perhaps surprisingly, it is not necessary for prior distributions to be proper probability distributions (i.e., for the area under the prior distribution to equal 1.0) in order for the posterior distribution to be proper. In fact, a well-known type of vague prior, called a Jeffreys prior, is often improper. The latter class of priors was suggested by the influential Bayesian pioneer Harold Jeffreys (e.g., see Jeffreys 1935). Jeffreys priors have the advantage of being invariant to transformation. This means that if $\pi(\theta)$ is the Jeffreys prior for some parameter θ, and ϕ is a one-to-one transformation $\phi = f(\theta)$, then the Jeffreys prior for ϕ is $\pi(\phi) = \pi(\theta) \left| \frac{d\phi}{d\theta} \right|$ (e.g., see Box and Tiao 1972). However, as mentioned above, a Jeffreys prior may be improper. For example, the Jeffreys prior for the variance of a normal distribution, σ^2, is given by $f(\sigma^2) \propto 1/\sigma^2$. This is improper; the area under the distribution from 0 to ∞ is infinite. Yet in concert with a normal likelihood and a suitable prior for the mean (μ), the resulting posterior distribution can be proper.

Although it is not uncommon for proper posterior distributions to arise from improper priors, it is not guaranteed. Hence extreme caution must be exercised whenever improper priors are used. Although sometimes useful, improper priors are used less frequently in current applied Bayesian analysis because modern software packages (e.g., OpenBUGS) generally require proper prior distributions.[1] Hence we usually find it more productive in practice to use vague proper prior distributions.

3.3   Conjugate Prior Distributions

When we have a prior distribution of a particular form and a likelihood which, when combined with the prior through Bayes theorem, yields a posterior of the same form

---

[1] OpenBUGS does permit the use of one improper prior distribution; a flat prior which is equal to a constant over an unbounded range. This is available using the dflat distribution.

as the prior, then the likelihood and prior are said to be conjugate.[2] Conjugate priors are convenient; advanced mathematics are not required to solve for the posterior distribution and the posterior is easily updated as new data arise (e.g., see Clark 2007). For example, if the sampling distribution is a binomial distribution with parameter p and the prior distribution for p is a beta distribution, then the posterior distribution is also a beta distribution. If a new, second data set was observed then the original posterior would become the prior for the second data set and the new posterior would also be a beta distribution. This particular model pair is often referred to as the beta-binomial model. Some other common conjugate model pairs are presented in Appendix A.

3.4 Prior Specification

3.4.1 Vague Priors

In the case of location parameters (e.g., the mean of a normal distribution), a flat or uniform prior over an appropriate finite interval is a conventional and convenient vague prior. Care must be taken to ensure that the bounds of the distribution do not exclude any intervals supported by the likelihood. For example, suppose we are estimating the carbon biomass of a forest stand. Furthermore, let's suppose we are satisfied with specifying a normal likelihood. Finally suppose we are certain the mean (μ) will lie between 50 and 80 tC/ha. We might then specify a uniform prior distribution for μ with bounds of 50 and 80, i.e., $\mu \sim \mathbf{Unif}(50, 80)$. This suggests that we are sure the μ lies between 50 and 80, but that all intervals of the same length within (50, 80) are equally likely. Of course, this also means that it is *impossible* for any interval outside (50, 80) to receive positive support in the posterior distribution, no matter what the likelihood reveals.

A common way to avoid the possibility of the prior excluding intervals supported by the likelihood is to instead specify a normal prior distribution for μ, but with a variance so large that the prior is nearly flat and effectively equivalent to a uniform distribution, e.g., $\mu \sim \mathbf{N}(0, 10000)$. This distribution gives a prior probability of 0.5 to negative values since it is centered at 0. One could truncate the normal prior to eliminate this feature, but such a vague prior would ordinarily be swamped by the data and not pose a problem. Following the suggestion of Spiegelhalter et al. (1996) we believe it is wise to check that the prior standard deviation is at least an order of magnitude greater than the posterior standard deviation.

For scale parameters, specifying vague priors is more problematic. For example, consider the variance of a normal likelihood, σ^2. As mentioned above, the Jeffreys prior is $f(\sigma^2) \propto 1/\sigma^2$. However, as also noted above, this is an improper prior and

---

[2]It is the posterior and prior which are of the same form, not the posterior and likelihood. It is easy to remember this when we consider that the likelihood deals with the distribution of the *data*, but the posterior and prior both describe distributions of the *parameters*.

cannot be used with Bayesian software packages such as OpenBUGS. A common method for dealing with this situation is to specify a proper prior for σ^2. Historically the solution has been to use an inverse-gamma prior, with the parameters of the prior chosen so as to result in a relatively flat distribution in all areas conceivably supported by the likelihood. However, Gelman (2006) has noted difficulties with this approach when using hierarchical models (see Sect. 4.4), in which the parameters of the prior distribution are *themselves* given prior distributions, usually denoted as hyperprior distributions. We advocate Gelman's advice of using uniform priors on standard deviations in lieu of inverse gamma distributions on variances in hyperpiors.

There are no automatic methods for specifying vague prior distributions for standard likelihoods, although much work has been performed and much progress made in this pursuit, e.g., see Berger (2006). For Bayesian methods to become as widespread as non-Bayesian methods, it might be necessary for some automatic method to become widely accepted. However, in another sense, this is an advantage. The fact that there are no automatic procedures requires scientists to think about their prior specifications. This is good in our view, as is anything which requires scientists to think more deeply about their analyses.

3.4.2 Informative Priors

Strict subjectivist Bayesians insist on using informative priors which completely characterize the scientist's beliefs prior to the experiment, and a large body of literature on prior elicitation exists, e.g., see Savage (1971), Kadane (1980), or Kadane and Wolfson (1998). While we might agree with this view on a philosophical level, in our experience scientists are generally too busy to devote enormous amounts of time to specification of priors for all the unknown parameters in every analysis they undertake. We believe that in most situations, the scientist will have reasonable expectations for the possible values of some parameters, and little information on others. In important problems, it might be worthwhile to perform a literature search for reported parameter values in similar studies, as in Green et al. (1999). Or it might be feasible to canvass experts in the field for their opinions on possible values (especially the smallest and largest expected values). If probabilities of various intervals can be elicited or estimated, then this information could be used to specify a prior distribution by, say, histogram smoothing as discussed in Congdon (2006). On the other hand, suppose historical data and/or reported values are available. A reasonable approach might be to select a probability distribution, such as one of those presented in Chap. 2 that approximates the histogram of the historical data.

3.4.3 Example: Poisson Sampling Model with Vague and Informative Priors

Suppose we are studying invasive species and are interested in the density of the exotic Japanese barberry (*Barberis thunbergii*) in deciduous forests in New Jersey.

Table 3.1 Counts of Japanese barberry plants on ten square meter plots, sum = 143

10	12
18	11
18	13
10	26
12	13

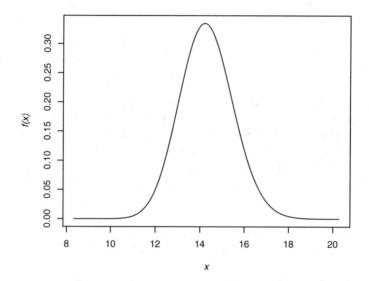

Fig. 3.1 Gamma posterior distribution with parameters 143.001 and 10.001

We randomly install $n = 10$ square meter plots in a forest and count the number of barberry plants in each (y_i) resulting in the data in Table 3.1.

Since the data are count data, a reasonable data model is the Poisson distribution, i.e., $Y \mid \lambda \sim \mathbf{Poi}(\lambda)$. From Appendix A, the conjugate prior for the Poisson data model is the gamma distribution. Hence if $\lambda \sim \mathbf{Ga}(\alpha_0, \beta_0)$ then $\lambda \mid Y \sim \mathbf{Ga}(\alpha_1, \beta_1)$, where $\alpha_1 = (\sum_i y_i + \alpha_0)$, $\beta_1 = (n + \beta_0)$, and n is the sample size. A reasonably vague prior density for λ is a gamma distribution with parameters 0.001 and 0.001 (Fig. 2.8). Combining this prior with a Poisson data model for the data in Table 3.1 yields a gamma posterior distribution with $\alpha_1 = 143.001$ and $\beta_1 = 10.001$. This distribution may be plotted in R using the code in Box 3.1. The first line in our R code is `rm(list=ls())`. This clears the R workspace and is a recommended best practice in R programming. The plotted posterior is displayed in Fig. 3.1.

Code box 3.1 R code to plot gamma distribution.

```
rm(list=ls())
alpha <- 143.001    # change to desired value
beta <- 10.001      # change to desired value
mean <- alpha/beta
sd <- sqrt(alpha/(beta^2))
lo_lim_x <- mean - 5 * sd
up_lim_x <- mean + 5 * sd
inc <- (up_lim_x - lo_lim_x)/10000
x <- seq(from = lo_lim_x, to = up_lim_x, by = inc)
ly <- (alpha-1)*log(x) + alpha*log(beta) - x*beta -
    lgamma(alpha)
y <- exp(ly)
plot(x, y, type ="l", xlab ="x", ylab="f(x)", xlim =
    c(lo_lim_x,up_lim_x))
```

One method to determine whether a prior distribution is informative or not is to compare the Bayesian estimate to the maximum likelihood estimate (Congdon 2006). If the two are similar, then the prior distribution exerts little influence on the estimate. In this case, the maximum likelihood estimate (MLE) is the sample mean, or $\hat{\lambda} = 14.3$ (see Johnson et al. 1993, among many others, for more information on maximum likelihood estimation and the Poisson distribution). The mean of the posterior distribution is $\alpha_1/\beta_1 = 14.299$, suggesting that the prior has very little influence.

Now suppose an additional 20 plots are installed in the same forest, resulting in the data in Table 3.2. In this case, it is sensible to use the posterior distribution from the initial study as an informative prior distribution for the second. The new, second posterior distribution, displayed in Fig. 3.2, has parameters $\alpha_2 = 413.001$ and $\beta_2 = 30.001$, where the subscript 2 indicates that these are the posterior parameters resulting from the second sample. The MLE for these 20 plots is 13.5, while the posterior mean is 13.8. Hence the informative prior did exert some influence over the estimate based on the second data set. Another commonly observed phenomenon is evident in Figs. 3.1 and 3.2. The posterior in Fig. 3.2 is concentrated much more tightly about the mean than the one in Fig. 3.1. This a result of the increased sample size, and illustrates how we become more sure about the posterior distribution as we observe more data.

Table 3.2 Counts of Japanese barberry plants on twenty square meter plots, sum $= 270$

13	17	19	7
11	18	14	11
8	17	12	15
14	11	15	10
15	17	16	10

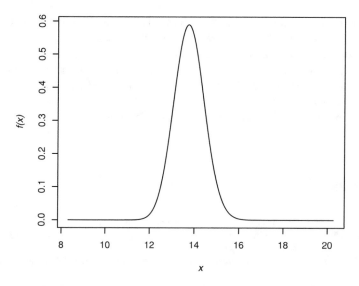

Fig. 3.2   Gamma posterior distribution with parameters 413.001 and 30.001

3.5   Exercises

1. Suppose we wish to estimate the number of tree seedlings in a forest. We randomly install ten square meter plots and count the number of seedlings in each resulting in counts of 51, 47, 55, 51, 57, 55, 44, 41, 53, and 56. Assume that the number of tree per plot follows a Poisson distribution.

   (a) Use a vague conjugate Gamma prior and derive the posterior distribution for the Poisson parameter λ. Plot the posterior distribution using the R code in Box 3.1.

   (b) Suppose we visit the same forest, randomly install 15 additional plots and observe the following number of seedlings per plot: 48, 52, 70, 38, 52, 46, 48, 48, 52, 59, 55, 56, 43, 41, and 56. Use the posterior distribution from part (a) as the prior distribution for this data and derive and plot the new posterior distribution for λ. How does this distribution compare to the one derived in part (a)? Note: The R script `plot 2 gamma distributions.R` is available online.

2. Suppose you want to estimate the percentage of clovers that have 4 leaves in a particular area. To that end, you collect a simple random sample of 20 plants and observe that two have 4 leaves.

   (a) Perform a Bayesian analysis, assuming you had no prior opinion regarding the percent of 4-leaf clovers in the area. Note: you need to specify a suitable likelihood and prior and specify parameters for the prior.

(b) Repeat the analysis on the same data set, but now assume that, prior to collecting data, you were pretty sure the percent of 4-leaf clovers was about 10%, and that your prior beliefs were represented by a **Be**(2,18) prior.

(c) Repeat part b, but suppose that, a-priori, you believed the percent of 4-leaf clovers was probably about 8%, and that your prior beliefs were represented by a **Be**(2,23) prior.

(d) Compare the priors and posteriors in (a), (b), and (c).

References

Berger, J. O. (2006). The case for objective Bayesian analysis. *Bayesian Analysis1*(3), 385–402.

Bernardo, J. M. (1979). Reference prior distributions for Bayesian inference. *Journal of the Royal Statistical Society: Series B*, *41*(2), 113–147.

Box, G. E. P., & Tiao, G. C. (1972). *Bayesian Inference in Statistical Analysis*. Reading: Addison-Wesley.

Clark, J. S. (2007). *Models for Ecological Data: An Introduction*. Princeton: Princeton University Press.

Congdon, P. (2006). *Bayesian Statistical Modeling* (2nd ed.). New York: Wiley.

Gelman, A., Carlin, J. B., Stern, H. B., Dunson, D. B., Vehtari, A., & Rubin, D. B. (2013). *Bayesian Data Analysis* (3rd ed.). New York: Chapman & Hall/CRC.

Gelman, A. (2006). Prior distributions for variance parameters in hierarchical models. *Bayesian Analysis*, *1*(3), 515–533.

Green, E. J., MacFarlane, D. W., Valentine, H. T., & Strawderman, W. E. (1999). Assessing uncertainty in a stand growth model by Bayesian synthesis. *Forest Science*, *45*, 528–538.

Jeffreys, H. (1935). Some tests of significance, treated by the theory of probability. *Mathematical Proceedings of the Cambridge Philosophical Society*, *31*(2), 203–222.

Johnson, N. L., Kotz, S., & Kemp, A. W. (1993). *Univariate Discrete Distributions* (2nd ed.). New York: Wiley.

Kadane, J. B. (1980). Predictive and structural methods for eliciting prior distributions. In A. Zellner (Ed.), *Bayesian Analysis in Econometrics and Statistics* (pp. 89–93). Amsterdam: North-Holland.

Kadane, J. B., & Wolfson, L. J. (1998). Experiences in elicitation. *The Statistician*, *47*, 3–19.

Savage, L. J. (1971). Elicitation of personal probabilities and expectations. *Journal of the American Statistical Association*, *66*, 783–801.

Spiegelhalter, D. J., Best, N. G., Gilks, W. R. & Inskip, H. (1996). Hepatitis B: A case study in MCMC methods. *Markov Chain Monte Carlo in Practice*, pp. 45–58.

Chapter 4
Elementary Bayesian Analyses

Before considering more advanced models which might be used in lieu of standard non-Bayesian approaches such as linear regression or Poisson regression, we start with some relatively simple Bayesian models. These will set the stage for the more sophisticated models covered in later chapters.

4.1 Beta-Binomial Model

In 2004, an unpublished study of bald eagle (*Haliaeetus leucocephalus*) nests was undertaken by the Endangered and Nongame Species Program (ENSP) of the New Jersey Department of Environmental Protection. Among other things, there was interest in estimating the probability of success of eagle nests, where success was defined as producing at least one eaglet during the year. A total of 41 nests were examined, of which 31 were successful. For ease of explanation, we will assume 30 successful nests were observed out of 40. A natural Bayesian model for this situation is the conjugate beta-binomial model (Appendix A). We assume that productive nests follow the binomial distribution:

$$f(X = x \mid n, p) = \binom{n}{x} p^x (1-p)^{n-x} = \frac{n!}{x!(n-x)!} p^x (1-p)^{n-x} \qquad (4.1.1)$$

where p is the probability of a productive nest, n is the number of nests examined and x is the number of productive nests found. In Appendix A, we see that the conjugate prior for p is the beta distribution:

$$f(p \mid \alpha_0, \beta_0) = \frac{\Gamma(\beta_0 + \alpha_0)}{\Gamma(\alpha_0) + \Gamma(\beta_0)} x^{\alpha_0 - 1} (1-x)^{\beta_0 - 1}, \quad 0 < p < 1; \ \alpha_0 > 0; \ \beta_0 > 0.$$

$$(4.1.2)$$

© Springer Nature Switzerland AG 2020
E. J. Green et al., *Introduction to Bayesian Methods in Ecology and Natural Resources*,
https://doi.org/10.1007/978-3-030-60750-0_4

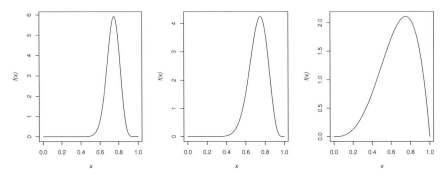

Fig. 4.1 Posterior distribution for the probability of a productive nest (p), given samples of size $n = 40, 20$, and 4

Since the beta distribution is a conjugate prior for the binomial likelihood, the posterior distribution for p is also a beta distribution, with parameters $(y + \alpha_0)$ and $(n - y + \beta_0)$, respectively (Appendix A). All that remains is to specify the parameters of the prior distribution for p. One reasonable approach is to set $\alpha_0 = 1$ and $\beta_0 = 1$. As shown in Fig. 2.9, this is equivalent to specifying a uniform prior on p, meaning that we are completely unsure about its value, except that it must lie in the interval $(0, 1)$. The posterior distribution for p is then **Be**(31, 11) and is displayed in Fig. 4.1. Also shown in Fig. 4.1 are the posterior distributions that would have obtained if had we observed 15 successful nests out of 20, or 3 successful nests out

Fig. 4.2 Informative prior distribution for p; a beta distribution with $\alpha = 5$ and $\beta = 2$

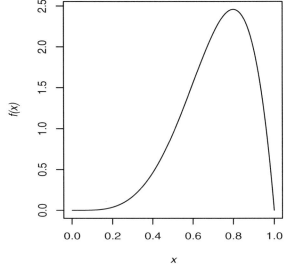

prior distribution for p

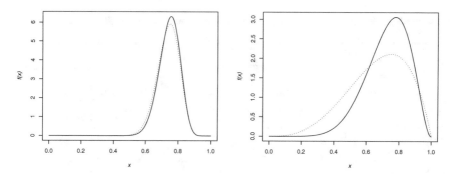

Fig. 4.3 Posterior distributions resulting from informative prior (solid line) and uniform prior (dotted line), for $n = 40$ and $n = 4$

of 4. Figure 4.1 demonstrates that, as would be expected, as n increases, our uncertainty over the value of p decreases and the posterior becomes more peaked with a reduced spread.

The prior distribution used above is based on the assumption that we have very little prior knowledge about possible values of p. In contrast, suppose we do have good prior knowledge, perhaps based on historical data. In particular suppose our prior distribution is **Be**(5, 2), as shown in Fig. 4.2. In Fig. 4.3 we overlay the posterior distributions resulting from using this prior and the posterior resulting from the uniform prior (**Be**(1, 1)), for the cases of 30 successful nests out of 40, and 3 successful nests out of 4, respectively. This figure displays a common result in Bayesian analyses; the posteriors resulting from very different priors may differ when data is scarce (as exemplified here by $n = 4$), yet the posteriors converge for larger data sets ($n = 40$ in this example).

4.2   Normal Model, Known Variance

ENSP has studied neo-tropical migrant birds for a number of years. As part of this study, red knots (*Calidris canutus rufa*) are captured in mist nets on selected New Jersey beaches in the spring of each year. Captured birds are measured, weighed, and banded. In this example, suppose we wish to estimate the mean weight (g) of red knots at a particular beach on a particular day in 2004. Thirty-seven birds were captured on the day in question; the data are in Table 4.1.

In order to ease into the study of normal models, we first assume that the population variance is known to be 370 g^2. We recognize that this is unrealistic; it is hard to envision a situation where the variance would be known while the mean was not. We make this assumption simply as a device to begin the study of normal models, and will relax it in the next section. A normal q-q plot, (e.g., see Wilk and Gnanadesikan 1968) of the 37 data points is shown in Fig. 4.4. If the data are normally distributed,

Table 4.1 Red knots weights in grams; $n = 37$, sample mean $= 144.1351$, sample variance $= 369.5646$

167	158	140	134	133
159	159	170	160	121
118	140	127	164	142
151	133	128	180	111
141	159	156	147	130
128	107	170	146	166
145	128	130	180	125
159	121	–	–	–

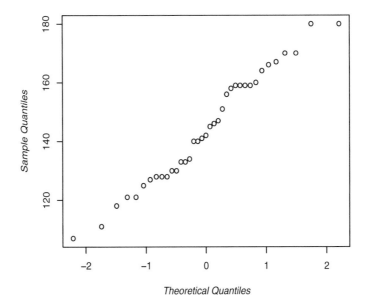

Fig. 4.4 Q-Q plot of red knot weight data

the points should fall approximately on a straight line. While there are some hints of non-normality in the figure, the data do appear to be roughly normal. Assuming the data are normally distributed, the likelihood for the observed data is:

$$L(\mu \mid \mathbf{y}) = (2\pi)^{-n/2}\, \sigma^{-n}\, e^{-\frac{1}{2}\sum_{i=1}^{n}\frac{(y_i-\mu)^2}{\sigma^2}}, \tag{4.2.1}$$

where $n = 37$ and $\sigma^2 = 370$.

From Appendix A, we see that the conjugate prior for the normal likelihood with known variance is a normal distribution. If the prior distribution is $\mu \sim \mathbf{N}(\mu_0, \sigma_0^2)$, then the posterior distribution is $\mu \mid \mathbf{y} \sim \mathbf{N}(\mu_n, \sigma_n^2)$, with

Fig. 4.5 Two prior distributions for mean of red knot data. Solid line: N(0, 10000); Dashed line: N(150, 100)

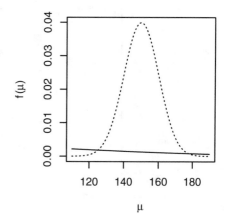

$$\sigma_n^2 = \frac{\sigma^2 \sigma_0^2}{\sigma^2 + n\sigma_0^2}, \text{ and } \mu_n = \sigma_n^2 (\frac{\mu_0}{\sigma_0^2} + \frac{n\bar{y}}{\sigma^2}),$$

where $\bar{y}$ is the sample mean ($n^{-1} \sum y_i$). Note that the posterior distribution behaves the way one would intuitively expect; as the prior variance σ_0^2 increases, more weight is attached to the sample mean, and as the variance of the observations (σ^2) increases, more weight is placed on the prior mean μ_0.

Suppose we either knew very little about potential weights of red knots, or wanted to let the data "speak for themselves" without specifying any strong prior beliefs. One way to do this would be to specify a normal prior with a variance so large that the prior is, for all intents and purposes, flat over all conceivable values for μ. For instance, we could adopt the prior $\mu \sim$ **N**(0, 10000). This prior is shown in Fig. 4.5. From 4.2, we compute the posterior mean and variance to be 144.0 and 10.0, respectively, i.e., $\mu \mid \mathbf{y} \sim$ **N**(144.0, 10.0), shown in Fig. 4.6.

On the other hand suppose that, based on past experience, we expected the mean weight of red knots on that particular day to be about 150 g, but we also thought it *could* be anywhere from 120 g to 180 g. The range of a normal distribution (the difference between the maximum and minimum values) is approximately equal to 6 standard deviations.[1] Hence in this case, $6\sigma = 60$, so $\sigma^2 = 100$, and a reasonable normal prior for mean red knot weight would be $\mu \sim$ **N**(150, 100). This prior is overlaid on the previous prior in Fig. 4.5. The resulting posterior distribution is

---

[1] If $X \sim$ **N**(μ, σ^2), then 99.8% of the values of X are in the interval $\mu \pm 3\sigma$.

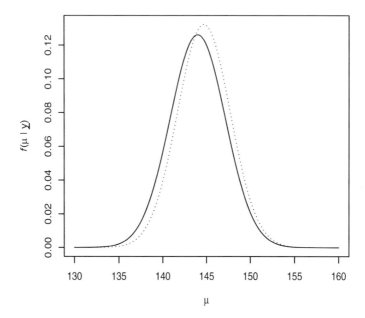

Fig. 4.6 Two posterior distributions for mean of red knot data, assuming known variance. Solid line: $\mathbf{N}(144, 10)$; Dashed line: $\mathbf{N}(144.7, 9.1)$

$\mu \mid y \sim \mathbf{N}(144.7, 9.1)$. In Fig. 4.6, we see this posterior overlaid on the previous one. Once again, even though the prior distributions in Fig. 4.5 are starkly different, the posterior distributions in Fig. 4.6 tell basically the same story.

4.3 Normal Model, Unknown Variance

The case of a normal data model with unknown variance is clearly more common and more realistic than the known variance case. It is also trickier; whether or not an analytical solution to the joint posterior is available depends on how the prior distributions are specified. Two forms suggest themselves[2]:

$$\mu \mid \sigma^2 \sim \mathbf{N}(\mu_0, \sigma^2/\kappa_0), \quad \sigma^2 \sim \mathbf{Ga}^{-1}(\alpha_0, \beta_0) \tag{4.3.1}$$

---

[2]Readers should not confuse these models with the normal data model with a non-conjugate flat Jeffreys prior for μ and a non-conjugate Jeffreys noninformative prior for σ^2. The posteriors for μ and σ^2 are available and well-known in the latter case (e.g., see pages 64-65 in Gelman et al. 2013). The latter model, while useful, does not lend itself well to hierarchical modeling and hence has been less widely applied in recent years.

or

$$\mu \sim \mathbf{N}(\mu_0, \sigma_0^2), \quad \sigma^2 \sim \mathbf{Ga}^{-1}(\alpha_0, \beta_0) \tag{4.3.2}$$

In 4.3.1, the prior distribution of μ is conditional on σ^2. As pointed out in Gelman et al. (2013) this prior often makes sense; it implies that our prior knowledge on μ is worth approximately κ_0 observations on Y. Under 4.3.1, analytic solutions to the joint posterior distribution for μ and σ^2 are available. In this case, it is the posterior for μ conditional on σ^2 and $\mathbf{y}$ that is conjugate with the conditional prior on μ. The marginal posterior on μ (i.e., $f(\mu \mid \mathbf{y})$) is *not* normal, but rather a t-distribution (the general form of the t distribution presented in 2.6.9).

In 4.3.2, μ and σ^2 are independent a-priori. This may seem like the more natural form for the normal data model, unknown mean and unknown variance problem. However, in this case analytic solutions to the posterior distributions are not available. Fortunately, approximation of the posteriors, to any desired level of precision, is readily obtainable using MCMC methods.

At this point we will drop the preceding assumption that we *know* the variance of the red knot weight data in Table 4.1. Furthermore, we will assume that (4.3.1) is an appropriate prior specification. As before, we will assume a prior mean for μ of 150 g. Suppose we feel this prior belief is worth about 3 observations (hence $\kappa_0 = 3$). To complete our prior specification, we will specify that, a-priori, σ^2 follows an inverse-gamma distribution with parameters $\alpha_0 = 0.001$ and $\beta_0 = 0.001^3$; this implies that the prior distribution for the precision, $1/\sigma^2$, is a gamma distribution with parameters $\alpha_0 = 0.001$ and $\beta_0 = 0.001$. The latter distribution is displayed in Fig. 2.8. The distribution is relatively flat, although it gives very high mass to very small values for the precision (or very high values for the variance). However, as long as the data yield no appreciable support for these extreme values, this will typically not be a problem in the posterior distribution.

As shown in many sources (e.g., Gelman et al. 2013), the marginal posterior distribution for μ is the three-parameter t distribution, $t_{\nu_n}(\mu_n, \sigma_n/\kappa_n)$, where

$$\bar{y} = n^{-1} \sum_{i=1}^{n} y_i,$$
$$s^2 = (n-1)^{-1} \sum_{i=1}^{n} (y_i - \bar{y})^2,$$
$$\kappa_n = \kappa_0 + n, \quad \nu_0 = 2\alpha_0,$$
$$\nu_n = \nu_0 + n, \quad \sigma_0^2 = \beta_0/\alpha_0,$$
$$\mu_n = \frac{\kappa_0}{\kappa_0 + n}\mu_0 + \frac{n}{\kappa_0 + n}\bar{y},$$
$$\sigma_n^2 = \frac{1}{\nu_0 + n}\left(\nu_0\sigma_0^2 + (n-1)s^2 + \frac{\kappa_0 n}{\kappa_0 + n}(\bar{y} - \mu_0)^2\right).$$

Letting $\mu_0 = 150$, $\alpha_0 = 0.001$, $\beta_0 = 0.001$, $\kappa_0 = 3$, $n = 37$, $\bar{y} = 144.1351$, and $s^2 = 369.5646$, we obtain $\mu_n = 144.575$, $\sigma_n^2 = 362.1366$, and $\sigma_n^2/\kappa_n = 9.053414$. From Sect. 2.6.4, we know that μ_n is the mean of the marginal posterior, and the variance is given by $(\nu_n/(\nu_n - 2))(\sigma_n^2/\kappa_n) = 9.57$. The mean and variance are very

---

[3]This is equivalent to specifying that σ^2 follows a scaled inverse-chi-squared distribution with $\nu = 0.002$ and scale parameter $\tau = 1$ (Gelman et al. 2013).

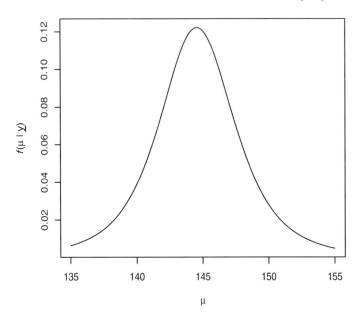

Fig. 4.7 Marginal posterior distribution for mean of red knot weight data, assuming unknown variance

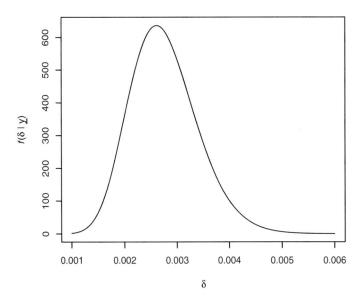

Fig. 4.8 Posterior distribution for precision $\delta = 1/\sigma^2$ of red knot weight data, assuming unknown variance

similar to the usual, non-Bayesian sample estimates ($\bar{y} = 144.1351$ and $s^2/n = 9.9882$, respectively). This is not surprising since we used vague priors on μ and σ^2. The marginal posterior distribution for μ is shown in Fig. 4.7.

The marginal posterior distribution for $1/\sigma^2$ is a gamma distribution with parameters $\alpha_n = \nu_n/2 = 18.501$ and $\beta_n = (\nu_n \sigma_n^2)/2 = 6699.89$ (e.g., see Gelman et al. 2013). The marginal posterior for $\delta = 1/\sigma^2$ is displayed in Fig. 4.8. It is evident in the figure that the marginal posterior no longer assigns substantial mass to extremely low values. Hence the data has overwhelmed that feature of the prior distribution for δ. From Sect. 2.6.5, we know the marginal posterior mean is $\alpha/\beta = 0.00276$ and the marginal posterior variance is $\alpha/\beta^2 = 4.1215 \times 10^{-7}$. The inverse of the posterior mean is thus $1/0.00276 = 362.3188$, which corresponds well with the usual non-Bayesian sample estimate $s^2 = 369.5464$.

4.4 Hierarchical Models

Hierarchical models are models in which the parameters in the prior distribution for the data model are themselves assigned prior distributions. These "priors for prior parameters" are conventionally referred to as hyperprior distributions (e.g., see Carlin and Louis 2009). For example, consider the classic rat growth model example in Gelfand et al. (1990). The data arose from a study in which the weights of 30 (presumably randomly selected) rats were monitored weekly for five weeks. Of course the growth of a rat probably follows a sigmoid pattern, but for the short duration of the study, the relationship of weight to age appeared linear (a common phenomenon for truncated growth series). The authors assumed homogenous variance at the data model stage, although they noted that heterogeneous variances could be easily accommodated. The data model[4] was thus

$$Y_{ij} \mid \alpha_i, \beta_i, \sigma^2 \sim N(\alpha_i + \beta_i x_{ij}, \sigma^2) \qquad (4.4.1)$$

where Y_{ij} is the weight of rat i at observation time j and x_{ij} is the age in days of rat i at observation time j; $j = 1, 2, \ldots, 5$ and $i = 1, 2, \ldots, 30$. Let $\theta_i = (\alpha_i, \beta_i)^\top$, where superscript $\top$ is the transpose operator. A Bayesian model is thus completed by specifying a prior distribution for θ_i and σ^2. Gelfand et al. (1990) assumed prior independence of θ_i and and σ^2, and let $\theta_i \sim \mathbf{MVN}(\mu, \Sigma)$ and $\sigma^2 \sim \mathbf{Ga}^{-1}(\nu/2, \nu\tau^2/2)$. However, instead of specifying values for μ and Σ, the authors assigned the following hyperpriors for them: $\mu \sim \mathbf{MVN}(\eta, \mathbf{C})$, and $\Sigma^{-1} \sim \mathbf{W}((\rho\mathbf{R})^{-1}, \rho)$.

Finally, specifying $\mathbf{C}^{-1} = \begin{pmatrix} 0 & 0 \\ 0 & 0 \end{pmatrix}$, $\nu = 0$, $\rho = 2$, and $\mathbf{R} = \begin{pmatrix} 100 & 0 \\ 0 & 0.1 \end{pmatrix}$ completed the model; the latter values were selected to provide vague prior information, and

---

[4]In Gelfand et al. (1990) there is a subscript "c" on the variance term and the prior and hyperprior parameters, probably indicating that this model is for rats in a control group. This is irrelevant for our purposes, so we have suppressed the subscript.

obviated the need to specify values for τ^2 and η. The model was fitted to the data with a Gibbs sampling algorithm.

The pertinent features of the rat growth model for the present discussion are the hyperpriors on μ and Σ. In the context of the study, it is clear that the growth curve parameters of individual rats were not of interest. Rather, interest focused on the mean growth curve for the population of rats from which these 30 animals were selected. The mean growth curve had hyperparameters μ and Σ and specifying hyperpriors on these allowed the authors to derive posterior distributions for μ and Σ (or, in the case of Gibbs (MCMC) sampling, generate samples from their marginal posterior distributions). All inferences then proceeded from these posterior distributions.

4.4.1   Random and Fixed Effects

In non-Bayesian modeling, it would be conventional to call the θ_i's in the preceding example random effects, implying that they parameterize the models for randomly chosen subjects. As above, when random effects are specified, interest usually centers on the parameters of the population from which the experimental subjects were randomly selected. In contrast, fixed effects models are useful when the experimental subjects are not randomly chosen, and when the models for specific individual subjects are important. For example, suppose one of the rats, say rat #5, in the above study had been fed a special diet. Then θ_5 might be of special interest. In this case it would be inappropriate to model rat 5 as if it was a randomly selected rat and θ_5 was a random realization from a larger population.

4.4.2   Exchangeability

In Bayesian modeling, the terms random effects and fixed effects are not often used, since even fixed effects are not "fixed" according to the Bayesian paradigm (see Sect. 1.3). However, Bayesian models often appeal to a related concept known as exchangeability. Suppose there are k experimental units, $y_i, i = 1, 2, \ldots, k$. Individual y_i's are said to be exchangeable if the probability density is invariant to permutations of the subscripts of the y_i's (e.g., see Gelman et al. 2013 or Bernardo and Smith 1994). In practical terms, this means that there is nothing distinctive about any of the individual y_i's, and all the subscript i does is separate data on one subject from another. In the rat example above, rats are considered to be exchangeable. The only difference among the weights from different rats is that they are from different rats. A-priori we have no reason to model the growth curve parameters of one rat differently from those of another rat.

In general, suppose we observe data from k groups and we believe the data $y_{ij}, i = 1, 2, \ldots, k; j = 1, 2, \ldots, n_i$, where n_i is the sample size from group i, are exchangeable. For instance, suppose the data are normally distributed with exchangeable means. Then the following might be a reasonable model:

$$Y_{ij} \mid \mu_i, \sigma^2 \sim \mathbf{N}(\mu_i, \sigma^2), \tag{4.4.2}$$

$$\mu_i \mid \theta, \tau^2 \sim \mathbf{N}(\theta, \tau^2), \quad \sigma^2 \sim \mathbf{Ga}^{-1}(0.001, 0.001), \tag{4.4.3}$$

$$\theta \sim \mathbf{N}(0, 10000), \quad \tau \sim \mathbf{Unif}(a, b). \tag{4.4.4}$$

The data model (4.4.2) specifies that the data from each group are normally distributed with group-specific means and a common variance (alternatively, we could specify group-specific variances).

The priors in (4.4.3) specify that the group means arise from a normal distribution with mean θ and variance τ^2, and the variance σ^2 follows an inverse-gamma distribution. The latter means that the inverse of the variance, or the precision, follows a gamma distribution. The parameters (0.001 and 0.001) are chosen so as to yield a flat or vague prior on the variance (see Fig. 2.8).

The hyperpriors in (4.4.4) are vague; the normal hyperprior on θ has a very large variance and the Uniform hyperprior on τ is flat over the range (a, b). In practice a and b would be assigned values to include all conceivable values of μ_i. In earlier work (prior to $\sim$ 1995) it was common to use gamma priors for precisions at all stages of the model hierarchy. However, Gelman (2006) demonstrated that this can be dangerous for higher level precisions, i.e., precisions of prior parameters or hyperprior parameters. This is because it's possible for higher level precisions to be very small (because there may be little information in the data on these parameters) in which case the spike in the Gamma(0.001, 0.001) distribution corresponding to small values can be problematic. Gelman (2006) recommended using uniform priors on higher level standard deviations, as we have done in (4.4.4), to avoid this.

It is interesting to note what happens when fitting a hierarchical model such as (4.4.2)–(4.4.4). θ is the mean of the group means, so the data from all the groups is used to estimate θ. On the other hand, μ_i is the mean for group i. Looked at on a single group level, μ_i is estimated as a weighted combination of the sample mean from group i and the overall mean θ. But the estimate for θ involves data from *all* the groups. Hence the data from all the other groups is used to estimate the mean for group i. This has the effect of anticipating the concept of "regression to the mean," whereby the group means are *shrunk* toward the overall mean (e.g., see Kelly and Price 2005).

4.4.3 Number of Levels in Hierarchical Models

In principle, there is no limit to the number of levels in a hierarchical model. For instance, in the rat example above, we could theoretically have specified "hyper-hyperpriors" on ν, τ^2, η, $\mathbf{C}$, ρ, and $\mathbf{R}$. The appropriate number of levels of prior distributions to specify varies from problem to problem. However, as mentioned in Carlin and Louis (2009), there is typically little advantage to adding additional priors beyond the second-stage prior, or what we called the hyperprior above. This is because the data are usually relatively non-informative for parameters above the hyperprior level and adding additional levels of prior structure generally results in only very small changes to the posterior distributions of the parameters at the data model or (first) prior distribution level.

Our experience suggests that, from a practical point of view, a reasonable number of hierarchical levels to specify is the *minimum* number which permits solution of the posterior distributions of the parameters of interest. For example, in the rat example the parameters of interest were μ and Σ. Since these were the parameters of the prior distribution, Gelfand et al. (1990) specified hyperpriors for these two parameters, and were thus able to solve for their posterior distributions.

4.5   Exercises

1. Suppose we observe the following random sample of weights (kg) of 15 deer:

$$50, 40, 43, 44, 45, 44, 44, 51, 50, 51, 52, 54, 39, 48, 43$$

   Fit the normal, unknown mean, unknown variance model in Sect. 4.3 to these data.
2. Use the OpenBUGS code normal model unknown mean and var.odc < (available online) to fit the the following model to the data in Exercise 1:

$$(y_i \mid \mu, \sigma^2) \sim \mathbf{N}(\mu, \sigma^2); \ i = 1, 2, \ldots, n;$$
$$\mu \sim \mathbf{N}(0, 10000); \ \sigma^2 \sim \mathbf{Ga}^{-1}(0.001, 0.001).$$

3. Compare the posterior distributions obtained in exercises 1 and 2.

References

Bernardo, J. M., & Smith, A. F. M. (1994). *Statistical Decision Theory and Bayesian Analysis*. New York: Wiley.

Carlin, B. P., & Louis, T. A. (2009). *Bayesian Methods for Data Analysis* (3rd ed.). Boca Raton: Chapman & Hall/CRC.

Gelfand, A. E., Hills, S. E., Racine-Poon, A., & Smith, A. F. M. (1990). Illustration of Bayesian inference in normal data models using Gibbs sampling. *Journal of the American Statistical Association, 85*(412), 972–985.

Gelman, A. (2006). Prior distributions for variance parameters in hierarchical models. *Bayesian Analysis, 1*(3), 515–533.

Gelman, A., Carlin, J. B., Stern, H. B., Dunson, D. B., Vehtari, A., & Rubin, D. B. (2013). *Bayesian Data Analysis* (3rd ed.). New York: Chapman & Hall/CRC.

Kelly, C. & Price, T. (2005). Correcting for regression to the mean in behavior and ecology. The American Naturalist 166(6), 700–707.

Wilk, M. B., & Gnanadesikan, R. (1968). Probability plotting methods for the analysis of data. *Biometrika*, *55*(1), 1–17.

Chapter 5
Hypothesis Testing and Model Choice

During the course of a scientific investigation, it is common to consider more than one model to explain or predict the phenomenon of interest. In some cases, it might be wise to cull the candidate models to a manageable number, and then use model averaging methods. However, often a scientist needs/wants to select a single model. In this chapter we consider the situation in which a scientist would like to select one of the candidate models to use for inference. Model selection shares many concepts with hypothesis testing, and so we begin this chapter with a discussion of hypothesis testing.

5.1 Examples

We will illustrate hypothesis testing/model choice with the aid of two data sets. The first consists of white tailed deer (*Odocoileus virginianus*) weights collected at two different times. This data set will be used for two hypothesis tests. The second consists of a set of fire scar intervals on red pine (*Pinus resinosa*) trees in Minnesota. These data will be used to choose between two models. The two data sets and the tests we will consider are described in the next two sections.

5.1.1 Deer Weights

Hunting of white-tailed deer was excluded from the Watchung Reservation in New Jersey for over 100 years (Green and Predl 2011). Due to increasing complaints about deer damage from nearby residents, hunting was permitted in 1994. Since then the deer herd has been culled annually. It may be of interest to determine if hunting pressure has exerted any influence on the condition of the deer herd. Green and Predl

© Springer Nature Switzerland AG 2020
E. J. Green et al., *Introduction to Bayesian Methods in Ecology and Natural Resources*,
https://doi.org/10.1007/978-3-030-60750-0_5

Table 5.1 Weight (kg) of 8 and 16 adult does harvested in 1994 and 2008, respectively

1994	2008	2008
37	41	43
35	48	43
46	41	42
48	43	50
37	43	50
41	31	55
44	35	33
51	42	39

examined the time trends in weight of male and female age classes, and reproductive rates of female age classes. Here we will simply test whether the mean weight of does harvested in 1994 was different from the mean weight of does harvested in 2008. It may also be of interest to compare the weight of harvested deer to published standards. Saunders (1988) reported that the mean weight of adult does in the Adirondack Mountains was about 155 lb, or 70 kg. Hence it might be interesting to test whether the mean weight of adult does in the Watchung Reservation in 2008 was equal to 70 kg (of course we recognize that the reservation is not actually in the Adirondack Mountains).

For the purposes of these examples, we will *assume* the harvested deer constitute random samples of the deer at the two times periods. There were 8 and 16 adult does harvested in 1994 and 2008, respectively (Table 5.1). Normal q-q plots Wilk and Gnanadesikan (1968) are shown for the weight data from each year (1994 and 2008) separately, and for both years combined in Fig. 5.1. The plots indicate that it is reasonable to assume the data are normally distributed. Let μ_1 and μ_2 denote the mean weights of adult does in 1994 and 2008, respectively. For ease of exposition, we will test the second hypothesis ($\mu_2 = 70$) first. Hence the first two null hypotheses we wish to test are

Test 1: $H_0 : \mu_2 = 70$, and
Test 2: $H_0 : \mu_1 = \mu_2$, or equivalently, $H_0 : D = 0$, where $D = \mu_1 - \mu_2$.

In order to test these null hypotheses, we must specify alternative hypotheses that we will accept if we cannot accept the null(s). Our alternatives are

Test 1: $H_A : \mu_2 \neq 70$, and
Test 2: $H_A : \mu_2 - \mu_1 \neq 0$ (or, equivalently, $D \neq 0$).

In classical hypothesis testing both of the above tests are called two-sided tests since in each the parameter of interest can differ from the hypothesized value in two directions. In Test 1, H_0 is false if either $\mu_2 > 70$ or $\mu_2 < 70$, and in Test 2, H_0 is false if either $D > 0$ or $D < 0$.

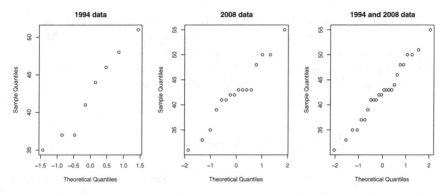

Fig. 5.1 Normal Q-Q plots for doe weight data from 1994, 2008, and both years combined

Table 5.2 Intervals in years between fire scars for red pine trees in Minnesota, from Clark (1990, 2007)

2	4	4	4	4	4	5	5	5	6	6	6	7	7
8	8	8	8	9	9	9	9	9	9	9	10	11	11
12	12	13	13	13	13	13	14	14	14	14	15	16	16
17	19	20	21	24	25	25	30	30	31	31	31	31	31
31	33	33	34	36	37	39	41	44	45	47	48	51	52
52	53	53	53	53	53	57	60	62	76	77	164	–	–

5.1.2 Fire Scar Intervals

The second data set will be used to choose between two potential sampling models. Clark (2007) presented data on fire scars in red pine trees in Minnesota. The data were originally reported in Clark (1990). The observations consisted of the intervals in years between observed fire scars. There were $n = 82$ observations. The data are reproduced in Table 5.2. Clark (2007) considered the question of whether these data were best fitted by an exponential or Weibull distribution. We will consider the same question.

The exponential density is given by

$$f(y \mid \lambda) = (1/\lambda)\, exp(-y/\lambda), \ y > 0, \ \lambda > 0, \tag{5.1.1}$$

and the Weibull density is[1]

$$f(y \mid v, \gamma) = v\gamma\, y^{(v-1)}\, exp(-\gamma y^v), \ y > 0, \ v > 0, \ \gamma > 0. \tag{5.1.2}$$

---

[1]This is the form of the Weibull density used in OpenBUGS. In R the Weibull is written in terms of a and b where $a = v$ and $b = \gamma^{-1/v}$. It is imperative that analysts check the definitions of probability densities used in statistical packages to avoid errors or nonsensical results.

It is immediately clear that the Weibull reduces to the exponential density if $\upsilon = 1$ and we let $\gamma^{-1} = \lambda$. Clark (2007) tested the null hypotheses $H_0 : \upsilon = 1$ versus the alternative $H_A : \upsilon \neq 1$ by calculating a 95% confidence interval for υ and found that it did not include 1, indicating that the Weibull density was the preferred model. He also performed a likelihood ratio test with the same result.

5.2   Hypothesis Testing Terminology

Hypothesis testing has its own terminology. Hypotheses are often referred to as point, simple, or composite. In the case of a *point* null, we are testing whether a parameter of interest takes on a particular hypothesized value (a certain point on the real number line). For example In Test 1, the null hypothesis $H_0 : \mu_2 = 70$ is a point null. Similarly, in Test 2, $H_0 : D = 0$ is a point null. Examples of non-point nulls, or range nulls, might be $H_0 : \mu_2 > 70$ or $H_0 : 0 \leq D \leq 5$.

A *simple* hypothesis is one in which the sampling model is completely specified. In our first example, weights of does would typically be modeled as realizations from a distribution with at least two parameters; one for the location and one for the scale of the distribution (e.g.., the mean and variance of a Normal distribution). Hence unless the variances are assumed to be known, the null hypotheses above are not simple because the scale parameters are unaccounted for in the hypotheses. A hypothesis in which the sampling distribution is not completely specified is called a *composite* hypothesis. The parameters not specified in the hypothesis are often called *nuisance* parameters. They must be accounted for in the analysis, even though they are presumably not of interest. In our doe weight example, the scale parameters are nuisance parameters.

5.3   Error Types and Acceptance/Rejection of Hypotheses

Statistical errors are commonly called Type 1 and Type 2. The former occurs when a true null hypothesis is incorrectly rejected, whereas the latter occurs when a false null hypothesis is incorrectly not rejected. The probabilities of Type 1 and 2 errors are customarily denoted by α and β, respectively. The standard hypothesis testing procedure is to *assume* the null hypothesis is true and then collect data and calculate a test statistic. If the latter assumes a surprising value, then the hypothesis is suspect. Given that the test has been performed correctly, we can usually control the probability of a Type 1 error, hence we can reject a null hypothesis with a stated α-level. Now, consider the situation where we cannot reject the null. Sometimes it is loosely stated that a null hypothesis is accepted. This is incorrect and should be avoided. As more data are observed, it is common for hypotheses that could not be rejected with few data to ultimately be rejected with more. For this reason, the proper scientific conclusions following a hypothesis test are (1) reject or (2) fail to reject. For more discussion of error rates, see Stauffer (2008), among many others.

5.4 Brief Philosophy of Hypothesis Testing

In general, most statistical methods for testing hypotheses roughly follow Popper's scientific method, in which hypotheses can never be proven to be correct, but can be disproven with empirical data (Popper 2002). However, rather than actually disproving a hypothesis based on empirical observations, modern statistical hypothesis testing involves evaluating whether the observed data are plausible or not, given the null hypothesis (or, in Bayesian testing, whether the hypothesis is plausible given the data). If it is deemed that the observations (or hypothesis) are (is) implausible, then it is assumed that either (i) an unlikely event has occurred, or (ii) the hypothesis is false. The usual procedure would then be to reject the null hypothesis, while accounting for the possibility of a false rejection. On the other hand, if the observations (or hypothesis) are (is) not implausible, then the usual procedure is to operate *as if* the hypothesis is true until new data come to light which lead to rejection of the hypothesis.

Alternative hypotheses like those in our Test 1 and Test 2 examples, $H_A : \mu_2 \neq 70$ and $H_A : (\mu_2 - \mu_1) \neq 0$, are sometimes called general, or non-specific hypotheses. They don't specify what the value of the parameter of interest is; they only specify what it isn't. There are two closely related classical schools of testing a point null against a general alternative hypothesis: the Fisher and Neyman–Pearson schools. In both of these, it is customary to calculate a test statistic and then compute the probability of observing a test statistic as extreme, or more extreme, than the one actually observed, under the assumption that the null hypothesis is true. This is called the p-value. In the Fisher school, one reports the p-value, and rejects the null hypothesis if the p-value is smaller than some pre-determined critical value (e.g.., 0.05). In the Neyman–Pearson school, one does not report the actual p-value, but only whether or not it was less than or greater to the critical value. In contrast to the Fisher and Neyman–Pearson methods, the standard Bayesian approach to testing a point null against a general hypothesis is to calculate the posterior probability that the null is true. If this is suitably small, then one rejects the null. Note that in the Fisher and Neyman–Pearson schools, one calculates the p-value by integrating over the sample space, which includes data which weren't actually observed (that is how the probability of observing a *more extreme* value for the test statistic is calculated).

The use of p-values has long been criticized and in 2016, the American Statistical Association was motivated to release a policy statement cautioning against their uncritical, routine use (Wasserstein and Lazar 2016). This was followed by an entire issue of *The American Statistician* (Volume 73 Number S1, March 2019) dedicated to removing p-valuse from their perch of arbiters of the significance of research results. We agree with much of the material in the ASA statement and the articles in the March 2019 *Am Stat* and recommend against the use of p-values (if a scientist insists on using p-values, then we recommend using 0.005 as the critical value, rather than 0.05. As detailed in Benjamin et al. (2018), this will lead to testing outcomes more in line with what practitioners expect when using $p = 0.05$).

We prefer formalizing hypotheses as statistical models and using model choice criteria to evaluate their plausibility. Rather than labeling results as "significant" or not, we prefer to indicate which hypothesis, the null or the alternative, is more compatible with the observed data. If effect sizes are deemed important, then we recommend computing Bayesian credible intervals.

In truth, we would prefer to determine which hypothesis is more compatible with the data via Bayes Factors, the traditional Bayesian method of discriminating between hypotheses and/or models. We present Bayes Factors in Sect. 5.6 and their attractiveness will become clear after reading it. However, computation of Bayes Factors is generally not easy so we do not recommend their routine use for Ecology or Natural Resource investigations.

5.5 Model Choice

Model choice is usually thought of as distinct from hypothesis testing, but in reality they are closely related. In fact, hypothesis testing can often be embedded in a model choice framework. For instance, in Test 2 if we are willing to assume that the observed weights are normally distributed, then our hypothesis test can equivalently be viewed as a choice between two models:

Model 1: $y_{ij} \sim \mathbf{N}(\mu, \sigma^2)$,
Model 2: $y_{ij} \sim \mathbf{N}(\psi, \sigma^2)$, $\psi = \mu_1 + I(\mu_2 - \mu_1)$,

where $i = 1, 2, \ j = 1, 2, \ldots, n_i$, n_i is the sample size for group i, $I = 0$ if $i = 1$ and $I = 1$ if $i = 2$. Model 1 specifies that the data all arise from the same normal distribution (i.e., the means for the two years are identical), whereas Model 2 specifies that the means for the two years are different.

5.5.1 Within-Sample Versus Out-of-Sample Prediction

When choosing a model, we frequently want to know how well it will perform when used for predictions on new data. For example, a wildlife manager may be interested in constructing a model to predict habitat suitability for a species of interest. The data for such a model might consist of covariates measured in a number of locations, some of which are deemed suitable and some which are deemed unsuitable. The manager may decide to use a general linear model (see Chap. 7) to predict suitability from the covariates. Presumably the ultimate objective is not to predict suitability for the sites already observed (we already know which of those are suitable and which are not), but to predict suitability for sites not included in the data. Hence the manager is interested in prediction for new or out-of-sample data, where "sample" refers to the data used to construct the model.

Now, suppose we agree that choosing model(s) on the basis of out-of-sample prediction accuracy is sensible. Then we are immediately confronted with a problem: Determination of out-of-sample prediction accuracy by definition depends on data we don't have. At this point, assuming we cannot collect more data, we have at least two possible solutions:

1. Randomly split the data into a fitting (or training) set and a validation set. Fit candidate models to the fitting data and use the fitted models to make predictions on the validation set. Then determine which model delivers the "best" predictions, or
2. Estimate out-of-sample predictive accuracy from within-sample predictive accuracy.

We prefer approach 1. We feel it is sensible to judge the ability of a model to make predictions on data to which it was not fitted by actually examining how well it predicts on data to which it was not fitted. This seems self-evident. However, there are drawbacks to this approach. First, a scientist needs to have sufficient data to get a reliable fit to the data in the fitting set (although this may be ameliorated by using leave-one-out methods; see Sect. 5.7.5). Secondly, this approach is typically a bit more computationally demanding, requiring more steps in the analysis. Finally, although the fitting data and the validation data are independent, there is some question as to how representative the validation data is of an entirely new data set. If the latter is collected at a different time and/or place or by different investigators, perhaps it will be different in some unknown respect than the independent validation set constructed by a random split of the existing sample data. We know of no way to account for this latter problem, but it is prudent to keep it in mind.

5.6 Bayes Factors

The traditional Bayesian method for testing hypotheses (or choosing between models) is to compute Bayes factors. The term "Bayes factor" appears to be due to Good (1958), but Jeffreys is normally credited with developing the method in, among other sources, Wrinch and Jeffreys (1921), Jeffreys (1935), and Theory of Probability, first published in 1939 (Jeffreys 1939/1961). Given data D, the Bayes factor (BF) in favor of H_0 over H_A, is

$$BF_{0A} = \frac{P(H_0 \mid D) \, / \, P(H_A \mid D)}{P(H_0) \, / \, P(H_A)}. \tag{5.6.1}$$

In a model choice context, the Bayes Factor in favor of model 1 (M_1) over model 2 (M_2) is:

$$BF_{12} = \frac{P(M_1 \mid D) \, / \, P(M_2 \mid D)}{P(M_1) \, / \, P(M_2)}. \tag{5.6.2}$$

For the remainder of this section, we will use the hypothesis formulation (BF_{0A}). The Bayes factor is a ratio of "odds." In general, the odds of an arbitrary event Q equals $[P(Q)/(1 - P(Q))]$, or $P(Q)/P(not\ Q)$. Suppose we are interested in whether it will rain tomorrow so we listen to evening news, on which the weather forecaster says that the probability of rain tomorrow is 30%. Then the odds of rain are 0.3/(1-0.3) = 0.43. Since the odds are less than 1.0, we would conclude that it probably isn't going to rain tomorrow (although our experience with the reliability of weather forecasts may lead us to carry an umbrella just in case).

In a test between null and alternative hypotheses, Q represents the null hypothesis, and *not* Q is the alternative. Hence the Bayes factor in (5.6.1) is the ratio of the posterior odds of H_0 to the prior odds of H_0. It measures how the odds in favor of the null hypothesis have changed as a result of the observed data. One might wonder why we wouldn't just use the numerator, i.e., the posterior odds of H_0. Perhaps it is easiest to see why we divide posterior odds by prior odds by way of an example: Suppose the posterior odds of a hypothesis were 1.2, but the prior odds were 1.4. Then, while the posterior odds favor the hypothesis, our faith in it has actually *decreased* in view of the data. This would seem to be something an investigator should know.

Interestingly, applying Bayes theorem to each term in the numerator in (5.6.1) yields

$$BF_{0A} = \frac{\left[\dfrac{P(D\,|\,H_0)P(H_0)}{P(D)} \right] \Big/ \left[\dfrac{P(D\,|\,H_A)P(H_A)}{P(D)} \right]}{P(H_0)\,/\,P(H_A)}$$

$$= \frac{P(D\,|\,H_0)}{P(D\,|\,H_A)}. \tag{5.6.3}$$

Hence the Bayes factor is the ratio of the probability of the data given H_0 to the probability of the data given H_A. This seems eminently logical and was long thought to be the only valid Bayesian way to compare hypotheses. In addition, from (5.6.1) it is immediately apparent that if the prior probabilities of the two hypotheses are equal, then

$$BF_{0A} = \frac{P(H_0\,|\,D)}{P(H_A\,|\,D)}, \tag{5.6.4}$$

i.e., the Bayes factor is the ratio of the posterior probability of H_0 to the posterior probability of H_A.

Unfortunately, despite their intuitive appeal, Bayes factors can be problematic. One difficulty is that the Bayes factor cannot be used for models which include improper priors on parameters. Consider the term $P(D\,|\,H_i)$ Let y represent the data with density f and let θ_i be the parameter (scalar or vector) of the model corresponding to H_i. Then $P(D\,|\,H_i) = p(y\,|\,H_i) = \int_{\theta_i} f(y\,|\,\theta_i, H_i)\,p(\theta_i)\,d\theta_i$. If

$p(\theta_i)$ is improper then it follows that $p(y \mid H_i)$ is also improper and hence the Bayes factor is not well defined, e.g., see Carlin and Louis (2009).

In addition, the numerator and denominator in (5.6.3) can be unstable and difficult to compute (Gelman et al. 2013). A number of methods of approximating Bayes factors have been advanced and many of these methods are discussed in detail by Kass and Raftery (1995) and Congdon (2006). From the scientist's viewpoint, a difficulty with this approach is that the method must be tailored to the problem. In theory this is fine and, for critical hypotheses, investing time to develop and code an algorithm would be reasonable. However, this approach may be unwieldy for routine hypothesis testing.

5.7 Information Theoretic Metrics

As mentioned earlier, hypothesis tests can often be re-written as model choice problems. When possible, we recommend translating null and alternative hypotheses into models and using model choice methods to discriminate between the two. Our preferred model choice method is to use information theoretic measures. Arguably, the most widely used information theoretic measure is Akaike's Information Criterion (*AIC*). *AIC* is *not* a Bayesian method, but it shares some features with Bayesian information theoretic methods and hence we discuss it in the next subsection, followed by discussion of several Bayesian alternatives.

5.7.1 AIC

It has become standard practice in Ecology and Natural Resource management to determine model adequacy or model "goodness" with Akaike's Information Criterion, better known by its acronym *AIC* (e.g., see Burnham and Anderson 2002). *AIC* is based on comparing the "information" in one model to that in another. This begs the question "What is information?". In an information criterion setting, supplying information can be regarded as reducing uncertainty, i.e.,

information supplied = prior uncertainty - posterior uncertainty

In a classic paper, Shannon (1948) proposed some basic axioms about the uncertainty regarding a random variable, say Y. He did this by describing prior knowledge about Y in terms of a probability distribution, $p_Y(y)$. Our prior uncertainty about values of Y is then a function of $p_Y(y)$, or more simply $p(y)$. Call this function $H(p)$.

Shannon presented three "common-sense" axioms which H should satisfy, and showed that as a result of these three axioms, H must be of the form

$$H(p) = -K \sum_y p(y) \, \log p(y), \tag{5.7.1}$$

where K is a constant.[2]

In another landmark paper, Kullback and Leibler (1951) used Shannon's notion of information to describe how to measure the information lost when a model g is used to approximate a "true" model f. This turned out to be:

$$I(f, g) = \int \log\left(f(y)\right) f(y) dy - \int \log\left(g(y \mid \theta)\right) f(y) dy$$
$$= E_f\left[\log(f(y))\right] - E_f\left[log(g(y \mid \theta))\right] \tag{5.7.2}$$

The second line follows from the first because of the technical definition of the expectation of a function, e.g., see Casella and Berger (2001). In the above, we haven't included parameters for f. We're assuming that f is the "truth." Parameters are a human construct; the "truth" has no parameters. On the other hand g is a model with parameter(s) θ.

This is an attractive construct, but not really useful by itself. It requires knowledge of f and θ. We ordinarily know neither (if we knew f, we wouldn't be trying to model it!). However, suppose we are interested in discriminating among a number of models. Observe that since f is the "truth," the first term on the RHS of (5.7.2) is the same for all models, and can be considered to be a constant, say C.

Next, let $\hat{\theta}$ be the maximum likelihood estimator (Sect. 1.3), and let the observed data be denoted by y. Akaike (1973) showed that $\log p(y \mid \hat{\theta})$ is a biased estimate of $E_f\left[\log(g(x \mid \theta))\right]$ and that the bias is asymptotically equal to k where k is the number of parameters in the model. Hence a reasonable estimator for $E_f\left[\log(g(y \mid \theta))\right]$ might be $\left[\log p(y \mid \hat{\theta}) - k\right]$. To be consistent with earlier statistical work, Akaike multiplied through by -2[3]:

$$AIC = -2 \log p(y \mid \hat{\theta}) + 2k \tag{5.7.3}$$

Now, note that the Kullback and Leibler (K-L) Information loss for a model, say model 1, is $(C - AIC_1)$ and the K-L information loss for another model, say model 2, is $(C - AIC_2)$, where AIC_i is the AIC for model i. Hence the difference is $(C - AIC_1) - (C - AIC_2) = AIC_2 - AIC_1$; the "truth" cancels out! Thus the procedure to follow is to calculate AIC for all candidate models. The model with the *lowest* AIC is the best (lowest is best because Akaike multiplied by -2). When k is large relative to the sample size n, then Burnham and Anderson (2002) showed that AIC should be modified to

$$AIC_c = -2 \log p(y \mid \hat{\theta}) + 2k + \frac{2k(k + 1)}{n - k - 1}, \tag{5.7.4}$$

---

[2]This is where the Shannon Diversity Index comes from: $H = -\sum_{i=1}^{S} p_i \log p_i$ where S is the total number of species and p_i is the frequency of the ith species, i.e., $p_i = n_i/N$, N is the total number of individuals of all species and n_i is the number of individuals of species i.

[3]There is a long history in Statistics of multiplying the log-likelihood, $\log p(y \mid \theta)$, by -2. This simplifies the expression when the density function under consideration is the normal density.

where the subscript c indicates "corrected." Actually, since AIC_c converges to AIC as n gets large, it's reasonable to simply always use AIC_c.

Information metrics such as AIC (and AIC_c) usually embrace Occam's razor (e.g., see Forster 2000 or Madigan and Raftery 2012), which roughly translates to wanting a model *just* complex enough to fit the data well. Occam's razor prizes parsimony and reminds us that the simplest explanation is usually the best. On the other hand, it is well known that model fit is guaranteed to improve (or at least not deteriorate) as the complexity of the model increases (e.g., see Draper and Smith 1998). Hence information metrics usually include a term for model fit and a penalty term for model complexity. In AIC these terms are $-2 \log p(y \mid \hat{\theta})$ and $(2k)$, respectively.

Several Bayesian information theoretic measures have been proposed in the literature. The most commonly encountered of these are the Bayesian Information Criterion (BIC), the Deviance Information Criterion (DIC), the Widely Applicable Information Criterion ($WAIC$), and the Leave-One-Out criterion (LOO). We discuss these in the next sections.

5.7.2 Bayesian Information Criterion

The Bayesian Information Criterion (BIC), also known as the Schwarz Criterion, was proposed by Schwarz (1978) in an attempt to define the marginal probability density of the data $p(y)$, given the model (e.g., see Gelman et al. 2013). As it happens, the BIC also can be used to provide an approximation to the Bayes Factor (Kass and Raftery 1995). BIC bears a striking resemblance to AIC $\left(BIC = -2 \log p(y \mid \hat{\theta}) + k \log(n) \right)$. However, AIC and several Bayesian methods (DIC, $WAIC$, and LOO) can be viewed as methods to estimate how well a model would perform on a new data set, whereas BIC was never intended by its developer to do that. Hence, we follow Gelman et al. (2013) and will not consider it further.

5.7.3 Deviance Information Criterion

The Deviance Information Criterion (DIC) is an analog to AIC. It was developed by Spiegelhalter et al. (2002). DIC has become a widely used measure of model fit in Bayesian analysis, and was reviewed by its developers after 12 years of use in Spiegelhalter et al. (2014). Almost certainly, a large part of the popularity of DIC among Bayesian practitioners is due its inclusion as an option under the Inference menu in WinBUGS and OpenBUGS.

As mentioned in Sect. 5.7.1, information metrics are generally composed of two parts: a measure of model fit and a penalty term for model complexity. For a model p with data y and parameter estimate $\tilde{\theta}$, it is common for the model *fit* to be measured by $-2 \log p(y \mid \tilde{\theta})$, and to be called the model *deviance* (see Spiegelhalter et al. 2002, among others).

As with AIC, Spiegelhalter et al. (2002) proposed that model complexity be measured by the number of parameters. However, as they pointed out, the number of parameters in a Bayesian model, particularly a Bayesian hierarchical model, is not straightforward.

Consider a model with parameter θ and data y. Let the density of y under the assumed model be $p(y \mid \theta)$. Then the marginal density of the data is

$$p(y) = \int_\theta p(y \mid \theta) p(\theta) d\theta. \tag{5.7.5}$$

We might further parameterize the model by including a hyperprior with parameter δ:

$$p(y, \theta, \delta) = p(y \mid \theta) p(\theta \mid \delta) p(\delta). \tag{5.7.6}$$

Then, depending on whether we are primarily interested in θ or δ, we might consider the "model" to be $p(y \mid \theta)$ with prior $p(\theta) = \int_\delta p(\theta \mid \delta) p(\delta) d\delta$ or $p(y \mid \delta) = \int_\theta p(y \mid \theta) p(\theta \mid \delta) d\theta$ with prior $p(\delta)$. Both approaches lead to the same marginal density of y, however they may be regarded as having different numbers of parameters, i.e., the dimension of θ and δ may differ.

Suppose we have available a sample of size n from the posterior distribution of θ, $\theta^{[i]}$, $i = 1, 2, \ldots, n$.[4] Spiegelhalter et al. (2002) estimated the *effective* number of parameters (p_D) as

$$p_D = \overline{D(\theta)} - D(\bar{\theta}), \tag{5.7.7}$$

where $\overline{D(\theta)}$ is the sample estimate of the posterior mean of the model deviance, i.e.,

$$\overline{D(\theta)} = -\frac{2}{n} \sum_{i=1}^{n} \log p(y \mid \theta^{[i]}), \tag{5.7.8}$$

and $D(\bar{\theta})$ is the model deviance measured at the mean of the posterior sample,

$$D(\bar{\theta}) = -2 \log p(y \mid \bar{\theta}), \tag{5.7.9}$$

$$\bar{\theta} = \sum_{i=1}^{n} \theta^{[i]}/n. \tag{5.7.10}$$

---

[4]MCMC methods deliver such samples.

Then Spiegelhalter et al. (2002) define DIC as the deviance at the posterior mean plus $2p_D$:

$$DIC = D(\bar{\theta}) + 2p_D \tag{5.7.11}$$

$$= D(\bar{\theta}) + 2\left(\overline{D(\theta)} - D(\bar{\theta})\right) \tag{5.7.12}$$

$$= 2\overline{D(\theta)} - D(\bar{\theta}) \tag{5.7.13}$$

$$= \overline{D(\theta)} + p_D. \tag{5.7.14}$$

Note that although the model size is a penalty, we *add* $2p_D$ to $D(\bar{\theta})$ in (5.7.11) because we are multiplying our expression of (model fit - model complexity) by -2 for historical reasons. The factor (-2) is already included in the deviance at the posterior mean.

When discriminating among a group of models, the one with the smallest DIC is estimated to be the model that would best predict an independent data set which has the same structure as the observed data.

It is difficult to say precisely what would constitute an important difference in DIC. As a rough guide, differences of more than 10 might definitely rule out the model with the higher DIC, differences between 5 and 10 are substantial, but if the difference in DIC is, say, less than 5, then it is reasonable to conclude that DIC does not distinguish between the competing models. This is especially important to keep in mind if the models lead to very different inferences.

There are at least two operational notes to keep in mind when using DIC. First, DIC can be negative because it is possible for probability densities to have values greater than 1.0 (as typically happens for distributions with ranges less than 1.0), and hence logarithms >0, resulting in a negative DIC after multiplying by -2. Second, it is also possible for p_D to be negative for certain types of likelihoods that are not log-concave with respect to the parameter(s).[5] In such situations, a negative p_D usually indicates that the posterior mode is a poor summary statistic for the posterior distribution, or that there is substantial conflict between the prior and the data (Spiegelhalter et al. 2002, p. 589).

The deviance used in the DIC metric delivered by **OpenBUGS** is is defined as $-2\log p(y \mid \theta)$, where y is the data, and θ comprises the "stochastic parents" of y, i.e., the parameters upon which the distribution of y depends, after collapsing over all logical relationships (Spiegelhalter 2014). It is important to bear this in mind because whether or not this is appropriate depends on the purpose of the investigation. For example, suppose the data are observed in groups and we have

---

[5]See Pratt (1981) for more on log-concave likelihoods.

$$y_{ij} \sim \mathbf{N}(\mu_i, \sigma_i^2),$$
$$\mu_i \sim \mathbf{N}(\psi, \tau^2), \ \sigma_i^2 \sim \mathbf{Ga}^{-1}(\alpha_1, \alpha_2)$$
$$\psi \sim \mathbf{N}(\phi, \delta^2), \ \tau \sim \mathbf{Unif}(a, b),$$
$$i = 1, \ldots, K; \ j = 1, \ldots, n_i,$$

where K is the number of groups and n_i is the sample size in group i. In most cases, $\alpha_1, \ \alpha_2, \ \phi, \ \delta^2, a$, and b will be specified to yield vague priors on $\sigma_i^2, \ \psi$, and τ.

For example, in a forestry context, the "groups" may be species, y_{ij} the biomass of tree j of species i, and μ_i the biomass Equation for species i.[6] In that case the future use would presumably be to predict biomass of a tree given its species and the appropriate likelihood would be $p(y_{ij} \mid \mu_i, \sigma_i^2)$. So, in a model choice exercise where the models have been fitted using an MCMC approach, this means we need to calculate the deviance of each observed y_{ij} given each set of values for $(\mu_1, \mu_2, \ldots, \mu_K, \sigma_1^2, \sigma_2^2, \ldots, \sigma_K^2)$ in the joint posterior sample. This is what is computed in OpenBUGS and hence the *DIC* option in OpenBUGS is appropriate.

In contrast, suppose we are interested in the average weights of red knots (*Calidris canutus*) when they land on New Jersey beaches during their annual Spring migration to the Arctic. To that end, a number of birds are captured and weighed on each of several beaches. Then y_{ij} may be the weight of bird j at beach i. In such a study, interest may be focused not on the mean and variance of bird weights on beach i (μ_i and σ_i^2), but instead on the average weight and variance of a bird in the population. This would be obtained by averaging over beaches and the appropriate likelihood would be $p(y \mid \psi, \tau^2) = \int_{\mu_i} \int_{\sigma_i^2} p(y_{ij} \mid \mu_i, \sigma_i^2) \times p(\mu_i, \sigma_i^2 \mid \psi, \tau^2) \, d\sigma_i^2 \, d\mu_i$. In a model choice exercise where the models have been fitted using an MCMC approach, this means we need to calculate the deviance of each observed y_{ij} given each value of μ_i, σ_i^2, ψ, and τ in the posterior sample. Unfortunately, this is *not* what OpenBUGS computes. However, users can output the joint posterior sample, read it into R, and compute *DIC* or use the R package **loo** and compute either *WAIC* or *LOO* (or both), using the proper likelihood. *WAIC* and *LOO* are discussed in the following sections.

5.7.4 Widely Applicable Information Criterion

The Widely Applicable Information Criterion (*WAIC*) (Watanabe 2010) is also known as the Watanabe-Akaike Information Criterion (e.g., see Gelman 2014; Spiegelhalter et al. 2014; or Vehtari et al. 2017). *WAIC* is similar in spirit to *DIC*, but instead of the deviance being conditioned on the posterior mean, it is averaged over the posterior distribution (a more "Bayesian" approach).

---

[6]The biomass equation would almost certainly involve covariates such as diameter and height which we suppress here for convenience.

As with *AIC* and *DIC*, *WAIC* is composed of two terms: one representing model fit, and a penalty for model complexity.

In *WAIC*, model fit is measured by the log pointwise predictive density (*lppd*):

$$lppd = \log \prod_{i=1}^{n} p_{post}(y_i) \tag{5.7.15}$$

$$= \sum_{i=1}^{n} \log \int_{\theta} p(y_i \mid \theta) p(\theta \mid y) d\theta, \tag{5.7.16}$$

where $p_{post}(y_i)$ is the posterior predictive distribution of y_i (see Sect. 5.9) and $p(\theta \mid y)$ is the posterior distribution of θ. As shown in Gelman (2014), given a sample from the joint posterior distribution (typically obtained from a MCMC routine), *lppd* can be computed as

$$lppd = \log \sum_{i=1}^{n} \left(\frac{1}{K} \sum_{j=1}^{K} p\left(y_i \mid \theta^{[j]}\right) \right), \tag{5.7.17}$$

where n is the sample size, i.e., the number of observations in the data, K is the size of the joint posterior sample, and $\theta^{[j]}$ is the jth value of θ in the joint posterior sample.

The rationale underlying using *lppd* as a measure of model fit is straightforward: for a given set of data points, we would prefer a model with a high posterior predictive density for those points. The posterior *predictive* density is used here as an indicator how a model would perform on a new, independent data set.

Gelman (2014) recommend the following penalty term for *WAIC*,[7] based on an estimate of the number of effective parameters in the model:

$$p_{waic} = \sum_{i=1}^{n} \text{var}_{post}\left(\log p(y_i \mid \theta) \right) \tag{5.7.18}$$

$$= \sum_{i=1}^{n} V_{j=1}^{K}\left(\log p\left(y_i \mid \theta^{[j]}\right) \right), \tag{5.7.19}$$

where var_{post} denotes the posterior variance and $V_{j=1}^{K}(x_j) = \frac{1}{K-1} \sum_{j=1}^{K}(x_j - \bar{x})^2$. Putting this all together (and multiplying by -2 for historical reasons) yields

$$WAIC = -2(lppd - p_{waic}) = -2\,lppd + 2\,p_{waic}. \tag{5.7.20}$$

Computing *WAIC* can be facilitated by using the R package **loo**. The user is required to input the original data and the log likelihood matrix. The latter is a

---

[7]Gelman (2014) present two penalty terms, but recommend p_{waic} for use based on its stability.

$K \times n$ matrix, where K is the joint posterior sample size and n is the data sample size. The value in cell (i, j) of the matrix is the log likelihood of data point j given the ith value of θ in the joint posterior sample.[8]

5.7.5 Leave-One-Out Criterion

The Leave-One-Out criterion (LOO) is explicitly intended to measure how the model under consideration would perform for a new data set. Given a sample of size n on the variable Y, LOO is defined as

$$LOO = -2 \sum_{i=1}^{n} \log p(y_i \mid y_{-i}), \qquad (5.7.21)$$

$$p(y_i \mid y_{-i}) = \int_{\theta} p(y_i \mid \theta) p(\theta \mid y_{-i}) d\theta, \qquad (5.7.22)$$

where y_{-i} represents all the y_i's except y_i. This makes sense; in (5.7.22) we see that the predictive density for y_i is based on all the observed data points except y_i. Hence y_i is playing the part of a new data point and y_{-i} is playing the part of the previously observed data. Of course there is nothing special about y_i, so in (5.7.21) we see that each data point gets to be "withdrawn" in turn.

While LOO has obvious appeal, there is an immediate problem: It appears that we would have to fit the model n times, withdrawing one data point each time. As a standard procedure, this would be unworkable. Fortunately, given a sample of size K from the joint posterior distribution, Vehtari et al. (2017) present the following estimate of LOO:

$$LOO = -2 \sum_{i=1}^{n} \log \left(\frac{\sum_{k=1}^{K} w_i^{[k]} p\left(y_i \mid \theta^{[k]}\right)}{\sum_{k=1}^{K} w_i^{[k]}} \right), \qquad (5.7.23)$$

where $\theta^{[k]}$ is the kth value for θ in the joint posterior sample, and $w_i^{[k]}$ is a Pareto-smoothed importance sampling weight (defined in Vehtari et al. 2017).

The R package **loo** can be called to compute LOO. As with $WAIC$, the user is required to input the original data and the $(K \times n)$ log likelihood matrix, where K is the joint posterior sample size and n is the data sample size. The value in cell (i, j) of the matrix is the log likelihood of data point j given the ith value of θ in the joint posterior sample.

---

[8]Alternatively, users can input a function to compute the log likelihood of a data point instead of the log likelihood matrix.

5.7.5.1 Application of DIC, WAIC and LOO to Doe Data

We can examine Tests 1 and 2 for the doe weight data by fitting models to the data and computing DIC, WAIC and/or LOO for each model. For Test 1 ($H_0 : \mu_2 = 70$ vs. $H_A : \mu_2 \neq 70$), we would fit two models: $i)$ $y_{2i} \sim \mathbf{N}(70, \sigma^2)$ and $ii)$ $y_{2i} \sim \mathbf{N}(\mu, \sigma^2)$. We would then compute DIC, WAIC, and/or LOO for each model and determine which model was best supported by the data. Readers are encouraged to do this on their own. Here we will examine the slightly more complicated Test 2: $H_0 : \mu_1 = \mu_2$ versus $H_A : \mu_1 \neq \mu_2$.

As noted earlier, the data appear to be normally distributed (Fig. 5.1). Since the null hypothesis did not specify anything about the variance, i.e., the variance is a nuisance parameter, we will fit four models:

(a) separate means per year, separate variances per year,
(b) separate means per year, constant variance,
(c) constant mean; separate variances per year,
(d) constant mean, constant variance.

The OpenBUGS code to fit all four models is shown in Box 5.1. We specify our standard vague priors on all unknown parameters ($\mathbf{N}(0, 1.0 \times 10^6)$ for means and $\mathbf{Ga}(0.001, 0.001)$ for precisions).

The code in Box 5.1 fits the model d. By commenting and uncommenting the appropriate lines of code, all four models may be fitted. Note that there is a for loop which must activated for models a–c. Also note the "nested index" we use for models a–c. For instance examine the statement

```
y[ i ] ~ dnorm(mu[yr[i]], tau[yr[i]]).
```

This is embedded in a loop in which i runs from 1 to N. So, for each value of i, we evaluate yr[i]. According to the vector yr, input in the data step, this will evaluate to a 1 or 2, depending on whether the ith observed weight was recorded in year 1 (1994) or 2 (2008). If it evaluates to a 1, then observation i follows a $\mathbf{N}(\mu_1, \sigma_1^2)$ density, whereas if it evaluates to a 2, observation i follows a $\mathbf{N}(\mu_2, \sigma_2^2)$ density. As always, bear in mind that in OpenBUGS, the normal density is parameterized in terms of the mean and precision.

Code box 5.1 OpenBUGS code for fitting models to doe weight data.

```
model
    {
        for( i in 1 : N ) {
#               y[ i ] ~ dnorm(mu[yr[i]], tau[yr[i]])   # model a
#               y[ i ] ~ dnorm( mu[yr[i]], tau )        # model b
#               y[ i ] ~ dnorm( mu, tau[yr[i]] )        # model c
                y[ i ] ~ dnorm( mu, tau)                # model d
        }
#       for( j in 1 : 2 ) {  # turn on loop for models a - c
#       mu[ j ] ~ dnorm(0.0,1.0E-6)      # model a or b
#       tau[ j ] ~ dgamma(0.001,0.001)   # model a or c
#       sigma[ j ] <- 1 / sqrt(tau[ j ])
#       } # end of loop for models a-c
```

Table 5.3 DIC, WAIC, and LOO for models a - d, doe weight data

Model	DIC	WAIC	LOO
a	161.7	160.9	161.2
b	159.5	159.3	159.4
c	159.4	158.8	159.0
d	157.3	157.1	157.0

```
          mu ~ dnorm(0.0,1.0E-6)          # model c or d
          tau ~ dgamma(0.001,0.001)       # model b or d
          sigma ← 1 / sqrt(tau)
     }

Inits
list(mu=c(0, 0), tau=c(1,1))   # model a
list(mu=c(0, 0), tau=1)        # model b
list(mu=0, tau=c(1,1))         # model c
list(mu=0, tau=1)              # model d

Data
list(y = c(37, 35, 46, 48, 37, 41, 44, 51,
           41, 48, 41, 43, 43, 31, 35, 42,
           43, 43, 42, 50, 50, 55, 33, 39),
     yr = c(1,1,1,1,1,1,1,1,
            2,2,2,2,2,2,2,2,
            2,2,2,2,2,2,2,2), N = 24 )
```

All four models converge very quickly. Since computation time was cheap, we ran each for 20,000 initial "burn-in" iterations and then computed DIC, WAIC, and LOO based on the subsequent 10,000 iterations. DIC has the advantage of being readily available from **OpenBUGS**. The DIC values for models a - d are displayed in Table 5.3.

WAIC and LOO are not computed within **OpenBUGS** but they may be computed in **R** using the package **loo**. In Box 5.2, we present the **R** code to do this. In this code box, as well as all others requiring **OpenBUGS** output to be read in, we include the **R** command `setwd(" ")`. The user must supply the path containing his/her data set inside the quotation marks.

Code box 5.2 **R** code for calculating WAIC and LOO for models a - d, doe weight data.
```
rm(list=ls())
# WAIC and LOO for models a - d, doe weight example
#  Be sure to set working directory!!
setwd(" ")
#  The package loo is used to compute LOO and WAIC
library("loo")
# input doe weights
y <- c(37, 35, 46, 48, 37, 41, 44, 51,
       41, 48, 41, 43, 43, 31, 35, 42,
       43, 43, 42, 50, 50, 55, 33, 39)
yr <- c(1,1,1,1,1,1,1,1,
        2,2,2,2,2,2,2,2,
```

```
          2,2,2,2,2,2,2,2)
Nobs <- length(y)
# read joint posterior samples
theta_a <- read.table("model
   a.out",header=FALSE,row.names = NULL)
theta_b <- read.table("model
   b.out",header=FALSE,row.names = NULL)
theta_c <- read.table("model
   c.out",header=FALSE,row.names = NULL)
theta_d <- read.table("model
   d.out",header=FALSE,row.names = NULL)
k <- 10000  # posterior sample size
# initialize parameter matrices and vectors
mu_a <- matrix(0,nrow=2,ncol=k); mu_c = rep(0,k)
mu_b <- matrix(0,nrow=2,ncol=k); mu_d = rep(0,k)
sig_a <- matrix(0,nrow=2,ncol=k); sig_b = rep(0,k)
sig_c <- matrix(0,nrow=2,ncol=k); sig_d = rep(0,k)
# copy posterior samples into appropriate matrices
   and vectors
mu_a[1,1:k] <- theta_a[1:k,2]
mu_a[2,1:k] <- theta_a[(k+1):(2*k),2]
sig_a[1,1:k] <- theta_a[((2*k)+1):(3*k),2]
sig_a[2,1:k] <- theta_a[((3*k)+1):(4*k),2]
mu_b[1,1:k] <- theta_b[1:k,2]
mu_b[2,1:k] <- theta_b[(k+1):(2*k),2]
sig_b[1:k] <- theta_b[((2*k)+1):(3*k),2]
mu_c[1:k] <- theta_c[1:k,2]
sig_c[1,1:k] <- theta_c[(k+1):(2*k),2]
sig_c[2,1:k] <- theta_c[((2*k)+1):(3*k),2]
mu_d[1:k] <- theta_d[1:k,2]
sig_d[1:k] <- theta_d[(k+1):(2*k),2]
# initialize log-likelihood matrices
log_lik_a <- matrix(0,nrow=k,ncol=Nobs)
log_lik_b <- matrix(0,nrow=k,ncol=Nobs)
log_lik_c <- matrix(0,nrow=k,ncol=Nobs)
log_lik_d <- matrix(0,nrow=k,ncol=Nobs)


# Compute log-likelihood for each weight and each set
   of
# parameters in joint posterior sample
for (i in 1:k){
  # parameters have index i
  for (j in 1:Nobs){
    # observations have index j
    log_lik_a[i,j] <- dnorm(y[j],mean = mu_a[yr[j],i],
        sd = sig_a[yr[j],i],log=TRUE)
    log_lik_b[i,j] <- dnorm(y[j],mean = mu_b[yr[j],i],
        sd = sig_b[i],log=TRUE)
    log_lik_c[i,j] <- dnorm(y[j],mean = mu_c[i], sd =
        sig_c[yr[j],i],log=TRUE)
    log_lik_d[i,j] <- dnorm(y[j],mean = mu_d[i], sd =
        sig_d[i],log=TRUE)
  }
}
```

```
# Get LOO and WAIC from loo
LOO_a <- loo(log_lik_a); WAIC_a = waic(log_lik_a)
LOO_b <- loo(log_lik_b); WAIC_b = waic(log_lik_b)
LOO_c <- loo(log_lik_c); WAIC_c = waic(log_lik_c)
LOO_d <- loo(log_lik_d); WAIC_d = waic(log_lik_d)
WAIC_a; WAIC_b; WAIC_c; WAIC_d
LOO_a; LOO_b; LOO_c; LOO_d
```

In addition to DIC, WAIC and LOO are also presented for each of the four models in Table 5.3. While the differences in each metric are not decisive, all four were lowest for model d. Hence we conclude that the data support the null hypothesis $H_0 : \mu_1 = \mu_2$, i.e., we fail to reject the null and will act as if it is true.

5.7.5.2   Application of DIC, WAIC and LOO to Fire Scar Data

The question regarding the fire scar data is whether the data better support an exponential model or a Weibull model. To that end, we fit each model using OpenBUGS and compute DIC for each. We also export the joint posterior distributions from OpenBUGS into R and compute WAIC and LOO.

Code box 5.3 OpenBUGS code to fit exponential and Weibull models to fire scar data.

```
MODEL 1 - Exponential
{
    for( i in 1 : N ) {
      y[i] ~ dexp(lambda)
    }
    lambda ~ dunif(0,10)
}
MODEL 2 - Weibull
{
    for( i in 1 : N ) {
      y[i]    ~ dweib(nu, gamma)
    }
    nu ~ dunif(0,10)
    gamma ~ dunif(0,10)
}
DATA
list(y=c(2,  4,  4,  4,  4,  4,  5,  5,  5,  6,  6,  6,  7,  7,  8,  8,
         8,  8,  9,  9,  9,  9,  9,  9,  9,  10, 11, 11, 12, 12,
         13, 13, 13, 13, 13, 14, 14, 14, 14, 15, 16, 16,
         17, 19, 20, 21, 24, 25, 25, 30, 30, 31, 31, 31,
         31, 31, 31, 33, 33, 34, 36, 37, 39, 41, 44, 45,
         47, 48, 51, 52, 52, 53, 53, 53, 53, 53, 57, 60,
         62, 76, 77, 164), N=82))
INITS
Weibull:
  list(nu=1, gamma=1)
Exponential:
  list(lambda=0.1)
```

Table 5.4 DIC, WAIC, and LOO for exponential and Weibull models, fire scar data

Model	DIC	WAIC	LOO
Exponential	700.8	699.5	700.7
Weibull	698.5	699.4	700.7

To fit, say, the exponential model using the code in Box 5.3, simply highlight the model section for the exponential model, and input the initial values for the exponential.

As evident in Box 5.3, we use vague priors for both models: $\lambda \sim$ **Unif**(0, 10) for the exponential model, and $\nu \sim$ **Unif**(0, 10), $\gamma \sim$ **Unif**(0, 10) for the Weibull. Both models converged very quickly, and computation time was fast, so we ran each for 30,000 iterations, discarded the first 20,000 and used the last 10,000 iterations for computation of DIC, WAIC, and LOO. DIC was available from **OpenBUGS** and is shown for each model in Table 5.4.

WAIC and LOO are not available directly from **OpenBUGS** but are easily computed using the R package **loo**. The R code to compute WAIC and LOO is presented in Box 5.4, and the resulting values are in Table 5.4. Neither DIC, WAIC, nor LOO is decisive; according to each statistic the two models (exponential and Weibull) are about equally good. This is contrast to the frequentist results obtained in Clark (2007) which favored the Weibull model.

Code box 5.4 R code to compute LOO and WAIC for exponential and Weibull models using fire scar data.

```
rm(list=ls())
# WAIC and LOO for exponential and Weibull models using fire
    scar data
# Be sure to set working directory!!
setwd(" ")
# The package loo is used to compute LOO and WAIC
library("loo")
# input fire scar data
y <- c(
  2, 4, 4, 4, 4, 4, 5, 5, 5, 6, 6, 6, 7, 7, 8, 8, 8,
  8, 9, 9, 9, 9, 9, 9, 9, 10, 11, 11, 12, 12, 13, 13,
  13, 13, 13, 14, 14, 14, 14, 15, 16, 16, 17, 19, 20,
  21, 24, 25, 25, 30, 30, 31, 31, 31, 31, 31, 31, 33,
  33, 34, 36, 37, 39, 41, 44, 45, 47, 48, 51, 52, 52,
  53, 53, 53, 53, 53, 57, 60, 62, 76, 77, 164)
Nobs <- length(y)
# read joint posterior samples
expon <- read.table("exp.out",header=FALSE,row.names =
    NULL)
weib <- read.table("Weibull.out",header=FALSE,row.names =
    NULL)
k <- 10000  # posterior sample size
# initialize parameter matrices and vectors
lambda <- rep(0,k)
nu <- rep(0,k)
gamma <- rep(0,k)
```

```
# copy posterior samples into appropriate matrices and
   vectors
  lambda[1:k] <- expon[1:k,2]
  gamma[1:k] <- weib[1:k,2]
  nu[1:k] <- weib[((k+1):(2*k)),2]
  rm(expon,weib)
# initialize log-likelihood matrices
  log_lik_e <- matrix(0,nrow=k,ncol=Nobs)
  log_lik_w <- matrix(0,nrow=k,ncol=Nobs)
# Compute log likelihood for each tree and each set of
# parameters in joint posterior sample
  for (i in 1:k){
  # parameters have index i
    for (j in 1:Nobs){
    # observations have index j
      log_lik_e[i,j] <- log(lambda[i]) - (y[j]*lambda[i])
      log_lik_w[i,j] <- log(nu[i]) +
          log(gamma[i])+((nu[i]-1)*log(y[j])) -
          (gamma[i]*(y[j]^nu[i]))
    }
  }
# Get LOO and WAIC from loo
  LOO_e <- loo(log_lik_e)
  WAIC_e <- waic(log_lik_e)
  LOO_w <- loo(log_lik_w)
  WAIC_w <- waic(log_lik_w)
  LOO_e; LOO_w
  WAIC_e; WAIC_w
```

5.8 Credible Intervals

Instead of calculating Bayes Factors for point null hypotheses or using model choice methods, an alternative approach is to evaluate the credibility of a hypothesized value by computing a credible interval and determining if the hypothesized value is inside the interval or not. A $(1 - \alpha)100\%$ credible interval for a parameter is one which includes, a-posteriori, $(1 - \alpha)100\%$ of the values of that parameter. Two desirable properties of credible intervals are (i) the density should be greater for every point in the interval than for every point outside it, and (ii) for a given probability content, the interval should be as short as possible. As pointed out by Box and Tiao (1972), "A moment's reflection will reveal that these two requirements are equivalent." For a symmetric unimodal distribution like the Normal, a $(1 - \alpha)100\%$ credible interval is derived by specifying that $(\alpha/2)100\%$ of the observations should be outside the interval in each tail. It is well known that, for the Normal distribution with mean θ and variance η^2, 95% of the observations are contained in the interval $(\theta - 1.96\eta, \ \theta + 1.96\eta)$ (see any mathematical statistics text, such as Wackerly et al. (2008) for more details on the Normal distribution).

Credible intervals are valid Bayesian posterior probability intervals. This is in contrast to non-Bayesian confidence intervals, which are not valid probability intervals after the intervals are constructed. It is easy to see why the latter is true: Suppose our confidence interval for some parameter θ is (θ_L, θ_U). Once an experiment has been performed and θ_L and θ_U have been computed, they are *known* constants. In the non-Bayesian view, θ is also an (unknown) constant. Since no constant can be between two others with a probability other than 0 or 1, a standard confidence interval cannot be a probability statement. For example, suppose we are studying the weight of red-tailed hawks (*Buteo jamaicensis*). Hence we collect a random sample of, say, 10 individuals, weigh each, and construct a standard 95% confidence interval. Further suppose that interval is (725 g, 1570 g). We may then ask ourselves "what is the probability that the *true* mean weight is between 725 and 1570 g?" The answer is 1 if it is and 0 if it isn't.

A 95% confidence interval is an interval established in such a way that, over repeated trials, if we constructed the interval the same manner each time, 95% of the intervals would be correct, i.e., would contain the true value of θ. Hence over repeated trials we are right 95% of the time, but on any particular trial (including the "trial" which resulted in the observed data) we are either right or wrong and we cannot know which. This problem does not arise in the Bayesian paradigm, where probability distributions are used to model our knowledge of the parameter. After collecting data, we are normally still somewhat uncertain over the value of the parameter, so Bayesian credible intervals, based on posterior distributions, are proper probability statements.

5.8.1 Point Null for Normal Mean, Variance Known

Consider our Test 1 from above, $H_0 : \mu_2 = 70$ versus $H_A : \mu_2 \neq 70$. At first, we assume the variance of the 2004 weight data (σ^2) is known, and equal to 40.0 (we will remove this assumption later). Let Y_2 denote weight of an adult doe in 2008. From Sect. 4.2 and Appendix A, we know that if we specify the prior $\mu_2 \sim N(\theta, \tau^2)$, then the posterior is $\mu_2 \mid Y_2 \sim N(\gamma, \kappa^2)$, where $\gamma = (\sigma^2 \theta + n_2\tau^2\bar{y}_2)/(\sigma^2 + n_2\tau^2)$; $\kappa^2 = \sigma^2\tau^2/(\sigma^2 + n_2\tau^2)$; n_2 is the sample size in 2008; and $\bar{y}_2$ is the usual sample mean, i.e., $\bar{y}_2 = \sum_i y_{2i}/n_2$. Observe that γ and κ^2 can be re-written as

$$\gamma = \frac{\frac{\theta}{\tau^2} + \frac{n_2\,\bar{y}_2}{\sigma^2}}{\frac{1}{\tau^2} + \frac{n_2}{\sigma^2}} \quad \text{and} \quad \kappa^2 = \frac{1}{\frac{1}{\tau^2} + \frac{n_2}{\sigma^2}}. \tag{5.8.1}$$

All that remains is to stipulate values for θ and τ^2. It would seem fair to use a vague prior, so let $\tau^2 = \infty$. Examination of (5.8.1) reveals that this obviates the need to specify a value for θ, and yields the posterior distribution $\mu_2 \mid Y \sim N(\bar{y}_2, \sigma^2/n_2)$. The sample mean of the 2008 weights (Table 5.1) was 42.4 and the sample size was $n_2 = 16$. Hence $\sigma^2/n_2 = 40/16 = 2.5$ and $\mu_2 \mid Y \sim N(42.4, 2.5)$. This posterior is displayed in Fig. 5.2. The credible interval for μ_2 is $42.4 \pm 1.96(2.5^{1/2})$ or (39.3, 45.5). Hence the reported value of 70 *kg* is an unreasonable value for the mean weight of does

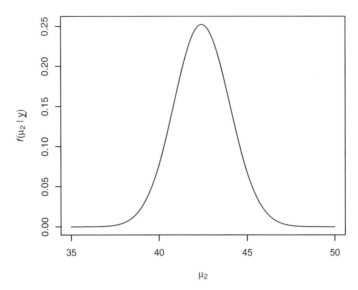

Fig. 5.2 Posterior density for mean 2008 doe weight with known variance and vague prior for the mean (μ_2)

harvested in Watchung Reservation in 2008. More study would be required to determine why. On the other hand, if the reported, hypothesized value had been 45 kg, this would not be unreasonable, and we would not reject the null (with Type 1 error rate (α) of 0.05).

5.8.2   Point Null for Normal Mean, Variance Unknown

As mentioned in Sect. 4.3, there are two commonly used prior specifications for the Normal mean, unknown variance case. In the first, (4.3.1), the prior distribution for the mean is conditional on the variance, while in the second, (4.3.2), the mean and variance are independent a-priori. While (4.3.2) may be more intuitive, (4.3.1) has the advantage that the posterior distribution is available analytically. Model (4.3.1) was illustrated in Sect. 4.3, so here we consider Model (4.3.2).

Suppose the model is

$$Y_2 \mid \mu_2, \sigma^2 \sim \mathbf{N}(\mu_2, \sigma^2), \quad \mu_2 \sim \mathbf{N}(\theta, \tau^{-1}), \quad \sigma^2 \sim \mathbf{Ga}^{-1}(\alpha, \beta). \qquad (5.8.2)$$

It is easy to generate arbitrarily large samples from the posteriors for μ_2 and σ^2 with an MCMC sampling package, such as OpenBUGS. Note that the prior variance of μ_2 is τ^{-1}; hence the prior precision, or inverse variance, of μ_2 is τ. We use this notation because the OpenBUGS software parameterizes the Normal density in terms of its mean and precision.

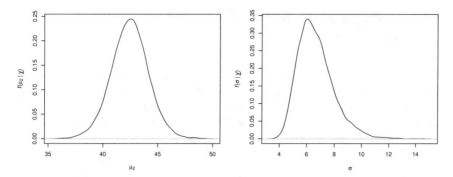

Fig. 5.3 Posterior densities for μ_2 and σ in doe weight example with unknown variance and vague priors on μ_2 and σ

In order to complete the model, we need to supply values for θ, τ, α, and β. We choose vague priors by specifying $\theta = 0$, $\tau = 1.0 \times 10^{-6}$, $\alpha = 0.001$, and $\beta = 0.001$. The **OpenBUGS** code is shown in Box 5.5. This model converges *very* quickly (e.g., the Geweke diagnostic in CODA indicated that convergence was obtained in < 1000 iterations). We ran the model for 50,000 iterations, and discarded the first 40,000. The resulting marginal posterior distributions for the mean (μ_2) and standard deviation (σ) are displayed in Fig. 5.3. The 2.5 and 97.5 percentiles of each distribution can be found in **OpenBUGS**, and are the approximate endpoints for 95% credible intervals. For μ_2, the interval is $(39.1, 45.8)$, while for σ the interval is $(4.7, 9.8)$. Hence, as before, the value 70 reported by Saunders (1988) is a highly unlikely value for μ_2.

Code box 5.5 OpenBUGS code for 2008 adult doe weights.

```
Model{
    for (i in 1:N){
        y[i] ~ dnorm(mu,tau)
        }
        mu ~ dnorm(0.0, 1.0E-6)
        tau ~ dgamma(0.001, 0.001)
        sigma ← 1/sqrt(tau)
}
Initial Values
    list(mu=0,tau=1)
Data
    list(y=c(41, 48, 41, 43, 43, 31, 35, 42,
             43, 43, 42, 50, 50, 55, 33, 39),
         N=16)
```

5.8.3 Testing Equality of Two Normal Means, Variances Unknown

The credible interval approach and MCMC sampling can also be used to test a null hypothesis of equality between two population means. In the remainder of this section, we discuss a simple method to use **OpenBUGS** to evaluate Test 2, $H_0 : (\mu_1 - \mu_2) = 0$ versus $H_A : (\mu_1 - \mu_2) \neq 0$, based on a point estimate of a contrast. A contrast among parameters is a linear function in which the coefficients of the parameters sum to 0. Hence in this case, the contrast is

$$(+1)(\mu_1) + (-1)(\mu_2).$$

Hence we are interested in the contrast $(\mu_1 - \mu_2)$, and specifically, we are interested in testing whether or not $(\mu_1 - \mu_2) = 0$, or in other words, whether 0 is a reasonable point estimate for $(\mu_1 - \mu_2)$.

As mentioned in Appendix A, a particularly useful feature of MCMC methods is that if we are interested in some function of parameter θ, say $g(\theta)$, where θ may be a vector, then we just need to calculate $g(\theta^i)$ on each iteration, where θ^i is the value of θ on iteration i. Following convergence of the chain, the sample of $g(\theta^i)$ values, $i = 1, 2, \ldots, n$, is a sample from the marginal posterior distribution of $g(\theta)$, where n is the posterior sample size (Gelfand et al. 1990). Thus if we compute the contrast of interest on each iteration, following convergence we will have a sample from the marginal posterior distribution of the contrast. We can then determine if the hypothesized value of the contrast (often 0) seems reasonable.

In the deer weight example the data from 1994 and 2008 have the same mean under H_0, whereas under H_A, the 1994 and 2008 means are different. For simplicity, we assume that the variances of the 1994 and the 2008 data are identical (it is straightforward to modify the **OpenBUGS** code for unequal variances). Let y_1 be the vector of observed 1994 weights and y_2 be the vector of observed 2008 weights. Then under the null hypothesis we have $y_1, y_2 \sim N(\mu, \sigma^2)$ where μ is the common mean for both years, whereas under the alternative hypothesis we have $y_1 \sim N(\mu_1, \sigma^2)$ and $y_2 \sim N(\mu_2, \sigma^2)$.

To complete the Bayesian model we require prior distributions for μ_1, μ_2, and σ^2. We will specify the vague priors $\mu_i \sim N(0, 1.0 \times 10^6)$, $i = 1, 2$, and $\sigma^2 \sim Ga^{-1}(0.001, 0.001)$. The analytic solution to the posterior distribution in this model (Normal data model, unknown mean, unknown variance, variance and mean a-priori independent) is mathematically intractable. Hence we use **OpenBUGS** to generate a sample from the joint posterior distribution. The **OpenBUGS** code is shown in Box 5.6. Here μ_1 and μ_2 are the mean weights for the 1994 and 2008 data, respectively; tau and sigma are the common precision and standard deviation, respectively; and diff is the contrast of interest, or $\mu_1 - \mu_2$.

Code box 5.6 Computing posterior distribution of a contrast among mean adult doe weights.

```
Model
    {
        for( i in 1 : N ) {
            y[i] ~ dnorm(mu[yr[i]], tau)
        }
        for( j in 1 : 2 ) {
            mu[j] ~ dnorm(0.0,1.0E-6)
        }
        diff ← mu[1] - mu[2]
        tau ~ dgamma(0.001,0.001)
        sigma ← 1 / sqrt(tau)
    }

Initial Values
 list(mu=c(0, 0), tau=1)

Data
 list(y = c(37, 35, 46, 48, 37, 41, 44, 51,
            41, 48, 41, 43, 43, 31, 35, 42,
            43, 43, 42, 50, 50, 55, 33, 39),
      yr = c(1,1,1,1,1,1,1,1,
             2,2,2,2,2,2,2,2,
             2,2,2,2,2,2,2,2),
      N = 24 )
```

The model in Box 5.6 converges quickly. We ran it for 50,000 initial burn-in iterations. The Geweke diagnostic in CODA indicated that after 5000 iterations, the sampler had converged. Since this model runs very quickly, we ran it for another 50,000 iterations to obtain our production sample. The marginal posterior densities for diff, μ_1, μ_2 and σ^2 are shown in Fig. 5.4. The posterior mean for diff was -0.07 and the 2.5 and 97.5 percentiles were -5.54 and 5.48, respectively. Thus an approximate 95% Bayesian credible interval for $(\mu_1 - \mu_2)$ is $(-5.54, 5.48)$. Clearly, 0 is a reasonable value for $(\mu_1 - \mu_2)$. The data support the null hypothesis that the mean weight of does in 1994 and 2008 was the same, at the $\alpha=0.05$ level.

5.9 Posterior Predictive Densities

One of the first requirements of a good model is that it be able to reproduce the data to which it was fitted. If it cannot do this reasonably well then there is little reason to suspect that it will predict new data with an acceptable degree of precision. We can assess a model's ability to reproduce the fitting data through posterior predictive distributions (PPD[s]). The theory underlying PPDs has been well-known for a long time (e.g., see Geisser 1993), and the widespread use of MCMC procedures has made computation of PPDs relatively easy (e.g., see Gelman and Hill 2007 or Gelman et al. 2013).

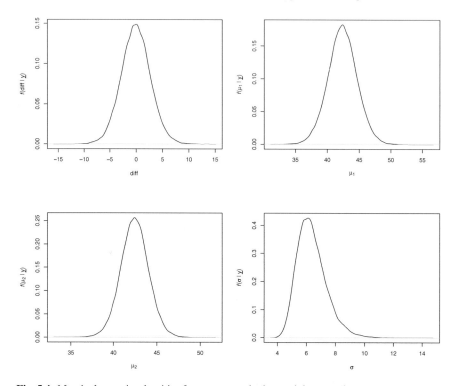

Fig. 5.4 Marginal posterior densities for parameters in deer weight example

The key to understanding the use of PPDs is to regard the likelihood as a sampling model. The sampling model defines the behavior of the data, given the parameter(s) θ. Bayes theorem shows that the data influence the posterior distribution *only* via the sampling model.

Imagine another replicate data set (y^{rep}) that *could* have been observed under the specified model. Since y^{rep} and y are two random outcomes given the model M with parameter(s) θ, we will consider them to be conditionally independent, given θ. This is akin to actually collecting a second data set and assuming it to be independent of the first. We need to find the density $p(y^{rep} \mid y)$ in order to generate replicate data sets, given the model and the observed data.

Note that

$$p(y^{rep} \mid y) = \int_{\theta} p(y^{rep}, \theta \mid y) \, d\theta, \qquad (5.9.1)$$

$$= \int_{\theta} p(y^{rep} \mid \theta, y) \, p(\theta \mid y) \, d\theta, \qquad (5.9.2)$$

$$= \int_{\theta} p(y^{rep} \mid \theta) \, p(\theta \mid y) \, d\theta. \qquad (5.9.3)$$

The equality in (5.9.1) is a standard integral calculus result. We move from (5.9.1) to (5.9.2) via the multiplicative rule of probability. Finally, we move from (5.9.2) to (5.9.3) by invoking the conditional independence of y^{rep} and y.

The density $p(y^{rep} \mid y)$ is the PPD. It allows us to generate replicate data sets that *could* have been observed but weren't, given the actual observed data and the model. The PPD can be used to generate a large number of replicate data sets. If the observed data looks like a sample from the set of replicate data sets, we can be confident the model is performing adequately. If not, the model is suspect. The use of PPDs will not necessarily identify the "optimal" model from a group of candidates, but it may eliminate models from further consideration. The computation and analysis of PPDs is a relatively new endeavor and is heavily reliant on modern computing power. In fact, it is fair to wonder what the inter-related fields of model choice and model checking would look like today had the development of modern computing power preceded the current body of statistical theory.

PPDs can be used to examine the relevant features of a model. When used in this capacity, they normally fall under the heading of *model checking*. If a number of models have been winnowed down to just a few remaining candidates (perhaps using information theory metrics such as WAIC or LOO, as discussed earlier in this Chapter), using PPDs to discriminate among them might fall under the heading of *model choice*. We are not proponents of categorizing the use of PPDs in either group; we believe practitioners do not particularly care what category a technique belongs to. They simply want a method to ensure that they are using a good model, and/or the best one out of a collection of possible models.

Generating a potential sample, y^{rep}, is relatively simple given the output from an MCMC sampling algorithm. The latter delivers observations from the joint posterior, $p(\theta \mid y)$. For every observation on θ in the joint posterior sample, we generate a sample from $p(y \mid \theta)$. In order to for the PPD to be directly comparable to the observed sample, we generate the same number of observations as there are in the observed data. If the model involves a covariate X then we expand (5.9.3) to $p(y^{rep} \mid y, x) = \int_\theta p(y^{rep} \mid \theta, x) \, p(\theta \mid y, x) \, d\theta$, where x denotes the observations on the covariate X.[9]

Suppose the joint posterior is of size m and we have n observations on y. Let $\theta^{[k]}$ be the kth value of θ in the joint posterior sample, $k = 1, 2, \ldots, m$. Then, for each $\theta^{[k]}$, generate n values from $p(y \mid \theta)$. These n values are one realization of y^{rep}; a potential alternative realization of the sample *if* the model is adequate. Doing this for each observation on θ in the posterior sample results in m potential samples. We can then examine whether the observed data looks like a sample from this collection of m samples.

We prefer to use graphical displays to determine whether the observed data y looks like a sample from the collection of potential replicate data sets ($y^{rep_i}, i = 1, 2, \ldots, m$). Any number of graphical procedures could be used to compare y and the generated y^{rep} data sets. If we want to examine certain features of the data, say

---

[9]The covariate might also be called a predictor variable or, in a regression context, the independent variable.

the 2.5 percentile, mean, median, or 97.5 percentile, this is easily accomplished by displaying the observed statistic on a histogram of the statistic computed from each of the m replicate data sets.

5.9.1 Fire Scar Data

We fitted the fire scar data with an exponential model and a Weibull model in **Open-BUGS**. Vague priors were used for all parameters; For the exponential density, we let $\lambda \sim$ **Unif**$(0, 10)$ and for the Weibull we let $\upsilon \sim$ **Unif**$(0, 10)$ and $\gamma \sim$ **Unif**$(0, 10)$. Computation was fast and cheap, so we ran each model for 50,000 iterations and discarded the first 25,000. The posterior sample for λ from the exponential model and the joint posterior sample for υ and γ for the Weibull model were exported using the CODA option in **OpenBUGS** and imported into R. For each observation in the posterior sample, we generated $n = 82$ observations from the corresponding distribution (exponential or Weibull). Then we computed histograms for the 2.5, 50, and 97.5 percentiles for the 25,000 replicate data sets and plotted the corresponding percentile from the observed data as a dark vertical bar on the histograms. The results are displayed in Fig. 5.5.

The 50.0 and 97.5 percentiles look reasonable for both densities (models). The observed 50.0 percentile looks a bit low compared to those from the PPDs for the Weibull, but not so unusual that it couldn't be accounted for by sampling error. On the other hand, the 2.5 percentiles appear to be problematic for both models. The models tend to generate 2.5 percentiles much lower than that in the observed data. A moment's reflection reveals why this may be occurring... both models are defined for $y > 0$, yet the very nature of the data reveals that it is impossible to observe a data point < 1. One potential remedy for this is to fit the data with a 3-parameter Weibull density:

$$f(y \mid \alpha, \upsilon, \gamma) = \upsilon\gamma \, (y - \alpha)^{(\upsilon-1)} \, exp(-\gamma(y - \alpha)^{\upsilon}), \quad y > \alpha, \ \upsilon > 0, \ \gamma > 0.$$
$$(5.9.4)$$

In (5.9.4), α is called the location parameter and y must be greater than α. We have written (5.9.4) to follow directly from the previous definition of the Weibull, (5.1.2). A more common definition of the 3-parameter Weibull is:

$$f(y \mid a, b, c) = \frac{c}{b} \left(\frac{y - a}{b}\right)^{c-1} exp\left(-\left(\frac{y - a}{b}\right)^{c}\right), \quad y > a, \ b > 0, \ c > 0.$$
$$(5.9.5)$$

In (5.9.5), the location parameter is a, and b and c are commonly called the scale and shape parameters, respectively. We fitted the fire scar data to the 3-parameter Weibull density. Unfortunately, **OpenBUGS** does not include this density as one its standard densities but it is relatively easy to specify a new distribution using the generic dloglik distribution. Our **OpenBUGS** code for this is in Box 5.7.

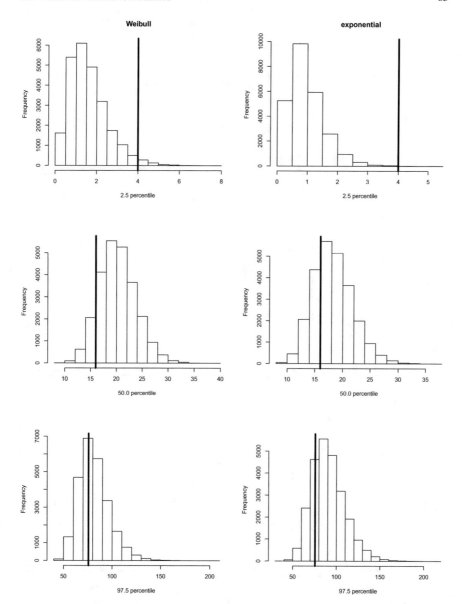

Fig. 5.5 Histograms for 2.5, 50.0, and 97.5 percentiles for exponential and Weibull densities from 25,000 replicate data sets

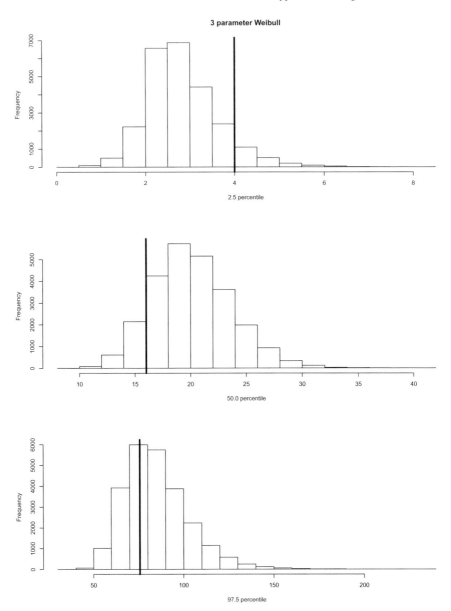

Fig. 5.6 Histograms for 2.5, 50.0, and 97.5 percentiles for 3 parameter Weibull density from 25,000 replicate data sets

Table 5.5 Bayesian p-values for 2.5, 50.0, and 97.5 percentiles for Weibull, exponential, and 3-parameter Weibull densities

Percentile	Weibull	Exponential	3-parameter Weibull
2.5	0.021	0.001	0.079
50.0	0.888	0.722	0.886
97.5	0.599	0.783	0.669

Code box 5.7 OpenBUGS code for generating values from a 3-parameter Weibull distribution, assuming the data are in the vector y and the sample size is N.

```
for( i in 1 : N ) {
    dummy[i] ← 0
    dummy[i] ~ dloglik(logLike[i])
    logLike[i] ← log(c) - log(b) +
    (c-1)*( log(y[i]-a) - log(b) ) - pow( ((y[i]-a)/b) ,
        c)
}
```

We fitted the 3-parameter Weibull density to the fire scar data in the same way in the same way we fitted the two parameter Weibull and the exponential densities. For prior distributions, we let $a \sim$ **Unif**$(0, 1.99999)$, $b \sim$ **Unif**$(0, 100)$, and $c \sim$ **Unif**$(0, 100)$. The resulting histograms for the 2.5, 50.0, and 97.5 percentiles from the PPDs for the 3-parameter Weibull are in Fig. 5.6. This model evidently does a better job at reproducing the 2.5 percentile then the 2-parameter Weibull or the exponential models, while doing about as well for the 50.0 and 97.5 percentiles. We can formalize this by computing the "Bayesian p-values" (Gelman et al. 2013).[10] For each model and percentile, we simply compute the proportion of percentiles from the PPDs which are less than or equal to the observed percentile. By convention, we would like these proportions to be between 0.025 and 0.975. For the fire scar data, the Bayesian p-values for the 2.5, 50.0 and 97.5 percentiles are presented in Table 5.5. The 3-parameter Weibull is the only model for which all three Bayesian p-values are in the range $(0.025, 0.975)$.

5.10 Exercises

The R library AICcmodavg contains the data set calcium. This data contains the calcium concentration in plasma of birds of both sexes following a hormone treatment. We reproduce the data here, ignoring sex. Group 1 birds did not receive the hormone whereas group 2 birds did:
Group 1: 16.5, 18.4, 12.7, 14.0, 12.8, 14.5, 11.0, 10.8, 14.3, 10.0
Group 2: 39.1, 26.2, 21.3, 35.8, 40.2, 32.0, 23.8, 28.8, 25.0, 29.3

---

[10]Given our reluctance to embrace the use of p-values for hypothesis testing, it may seem contradictory for us to advocate the use of Bayesian p-values. However, we use them simply as a guide to describe whether or not a model is able to adequately reproduce an important feature of the data. We are not interested in testing whether or not a given model is *true*. We know it is not; we are trying to determine if it will be useful.

1. Use **OpenBUGS** to fit the following normal, unknown means, unknown variances model to these data (here we are assuming group-specific variances):

$$y_{ij} \mid \mu_i, \sigma_i^2 \sim \mathbf{N}(\mu_i, \sigma_i^2)$$
$$\mu_i \sim \mathbf{N}(0, 10000); \quad \sigma_i^2 \sim \mathbf{Ga}^{-1}(0.001, 0001)$$
$$i = 1, 2; \quad j = 1, 2, \ldots, n_i$$

where n_i is the number of observations in group i.

Suppose we are interested in the following two null hypotheses:

(a) $H_0 : \mu_1 = \mu_2$ versus $H_A : \mu_1 \neq \mu_2$
(b) $H_0 : \sigma_1^2 = \sigma_2^2$ versus $H_A : \sigma_1^2 \neq \sigma_2$

2. Evaluate each null hypothesis using credible intervals.
3. Translate each null and alternative hypothesis into a model.

   (a) Fit the models with **OpenBUGS** and evaluate each using DIC, WAIC, and LOO.
   (b) Compute posterior predictive distributions for each model. Compare how well these correspond to the observed data.

References

Akaike, H. (1973) , Information theory as an extension of the maximum likelihood principle. In: Petrov, B., Csaki, F.(eds.), Second International Symposium on Information Theory. Akademiai Kiado, Budapest, Association for Computing Machinery, New York, NY (pp. 267–281).

Benjamin, D. J., Berger, J. O., Johannesson, M., Nosek, B., Wagenmaker, E. J., Berk, R. et al. (2018). Redefine statistical significance. *Nature Human Behavior*, *1*, 6–10.

Box, G. E. P., & Tiao, G. C. (1972). *Bayesian Inference in Statistical Analysis*. Reading, MA: Addison-Wesley.

Burnham, K. P., & Anderson, D. R. (2002). *Model Selection and Multi-Model Inference: A Practical Information-Theoretic Approach*. New York, NY: Springer.

Carlin, B. P., & Louis, T. A. (2009). *Bayesian Methods for Data Analysis* (3rd ed.). Boca Raton: Chapman & Hall/CRC.

Casella, G., & Berger, R. L. (2001). *Statistical Inference* (2nd ed.). Belmont, CA: Duxbury Press.

Clark, J. S. (1990). Fire and climate change during the last 750 years in northwestern minnesota. *Ecological Monographs*, *60*, 135–139.

Clark, J. S. (2007). *Models for Ecological Data: An Introduction*. Princeton, NJ: Princeton University Press.

Congdon, P. (2006). *Bayesian Statistical Modeling* (2nd ed.). New York, NY: Wiley.

Draper, N. R., & Smith, H. (1998). *Applied Regression Analysis* (3rd ed.). New York, NY: Wiley.

Forster, M. R. (2000). Key concepts in model selection: Performance and generalizability. *Journal of Mathematical Psychology*, *44*(1), 205–231.

Geisser, S. (1993). *Predictive inference*. New York: Routledge.

Gelfand, A. E., Hills, S. E., Racine-Poon, A., & Smith, A. F. M. (1990). Illustration of Bayesian inference in normal data models using Gibbs sampling. *Journal of the American Statistical Association*, *85*(412), 972–985.

Gelman, A., & Hill, J. (2007). *Data Analysis using Regression and Multilevel/Hierarchical Models*. New York, NY: Cambridge University Press.

Gelman, A., Carlin, J. B., Stern, H. B., Dunson, D. B., Vehtari, A., & Rubin, D. B. (2013). *Bayesian Data Analysis* (3rd ed.). New York: Chapman & Hall/CRC.

Gelman, A., Hwang, J., & Vehtari, A. (2014). Understanding predictive information criteria for Bayesian models. *Statistics and Computing, 24,* 997–1016.

Good, I. J. (1958). Significance tests in parallel and in series. *Journal of the American Statistical Association, 53,* 799–813.

Green, E. J. & Predl, S. (2011). Bayesian analysis of deer reproductive condition, *Contemporary Developments in Bayesian Analysis and Statistical Decision Theory: A Festschrift for William E. Strawderman, IMS Collection, Vol. 8 .*

Jeffreys, H. (1935). Some tests of significance, treated by the theory of probability. *Mathematical Proceedings of the Cambridge Philosophical Society, 31*(2), 203–222.

Jeffreys, H. (1939/1961). *Theory of Probability.* Oxford: University Press. Third edition in 1961, Oxford: University Press.

Kass, R. E., & Raftery, A. E. (1995). Bayes factors. *Journal of the American Statistical Association, 90*(430), 773–795.

Kullback, S., & Leibler, R. A. (1951). On information and sufficiency. *The Annals of Mathematical Statistics, 22,* 79–86.

Madigan, D., & Raftery, A. E. (2012). Model selection and accounting for model uncertainty in graphical models using Occam's window. *Journal of the American Statistical Association, 89,* 1535–1546.

Popper, K. (2002). *The Logic of Scientific Discovery.* New York: Routledge.

Pratt, J. W. (1981). Concavity of the log likelihood. *Journal of the American Statistical Association, 76*(373), 103–106.

Saunders, D. A. (1988). Adirondack mammals, Technical report, State University of New York, College of Environmental Science and Forestry.

Schwarz, G. (1978). Estimating the dimesnsion of a model. *The Annals Statistics, 6,* 461–464.

Shannon, C. E. (1948). A mathematical theory of communication. *The Bell System Technical Journal, 27,* 379–423.

Spiegelhalter, D. J., Thomas, A., Best, N. and Lunn, D. (2014). OpenBUGS User Manual, Version 3.2.3. http://www.openbugs.net/Manuals/Manual.html.

Spiegelhalter, D. J., Best, N. G., Carlin, B. P., & Van der Linde, A. (2002). Bayesian measures of model complexity and fit (with discussion). *JRSS B, 64*(4), 583–616.

Spiegelhalter, D. J., Best, N. G., Carlin, B. P., & Van der Linde, A. (2014). The deviance information criterion: 12 years on. *JRSS B, 76*(3), 485–993.

Stauffer, H. B. (2008). *Contemporary Bayesian and Frequentist Statistical Research Methods for Natural Resource Scientists.* New York: Wiley.

Vehtari, A., Gelman, A., & J., G., (2017). Practical Bayesian model evaluation using leave-one-out cross-validation and WAIC. *Statistics and Computing, 27,* 1413–1432.

Wackerly, D., Mendenhall, W., & Scheaffer, R. L. (2008). *Mathematical Statistics with Applications* (7th ed.). New York, NY: Duxbury Press.

Watanabe, S. (2010). Asymptotic equivalence of Bayes cross-validation and widely applicable information criterion in singular learning theory. *Journal of Machine Learning, 11,* 3571–3594.

Wilk, M. B., & Gnanadesikan, R. (1968). Probability plotting methods for the analysis of data. *Biometrika, 55*(1), 1–17.

Wrinch, D., & Jeffreys, H. (1921). On certain fundamental principles of scientific inquiry. *Philosophical Magazine, 42,* 369–390.

Chapter 6
Linear Models

One of the most widely used statistical models is the ubiquitous linear model. It arises in regression contexts, where the object is to relate a variable of interest to another variable, in experimental design contexts where an observation is modeled as a function of variable(s) that represent the experimental design, and in an analysis of covariance context which is a mix of the previous two settings. It also arises in less obvious ways, e.g., in general linear models (GLM[s]).

We start with a bit of terminology. Consider two variables X and Y, which we suspect are linearly related. Suppose our ultimate interest is in Y, but we suspect that X may be related to Y. Knowledge of X may help reduce the expected variability in Y or may help in predicting the value of Y for a new observation. To that end, we gather a sample of size n in which we record the values of X and Y for each sample unit, i.e., we gather n pairs of (X, Y). The linear model is then

$$y_i = \alpha + \beta x_i + e_i, \ i = 1, 2, \ldots, n, \tag{6.0.1}$$

where y_i and x_i are the observations on Y and X for sample unit i, and e_i is the unobserved random error. Equation 6.0.1 is sometimes loosely written without the error term. However, since the equation is a strict equality, that practice should be avoided.

The model in Eq. 6.0.1 is often called a *simple* linear model. This term is relative; the model may not be "simple" in an absolute sense, but it is arguably *simpler* than other forms of the linear model.

In Eq. 6.0.1, α is referred to as the intercept and β as the slope. The variable on the LHS of the equation (Y) is usually called the dependent, response, or outcome variable, and the variable on the RHS is called the independent, regressor, covariate or design variable. In a regression context, the terms dependent and independent variables are arguably the most common labels. This is somewhat unfortunate since the use of the terms might lead naive users to conclude that Y actually "depends" on X, i.e., that there is a causal relationship. This may be true, or it may not. Often, Y and

© Springer Nature Switzerland AG 2020

E. J. Green et al., *Introduction to Bayesian Methods in Ecology and Natural Resources*,
https://doi.org/10.1007/978-3-030-60750-0_6

X both depend on other factors, in which case the terms dependent and independent variable are best considered simply as placeholders instructing us on which side of an equation a variable belongs. For example, consider a model in which the dependent variable is individual tree volume and the independent variable is individual tree diameter. These two variables may be related linearly (perhaps after transformation of one or both variables), but both "depend" on a host of other factors, such as age, site quality and inter-tree competition.

A crucial assumption with respect to linear models concerns the *conditional* variance of the dependent variable. This is the variance of Y *given*, or *conditional* on, X and is usually denoted by $\sigma^2_{y.x}$. If $\sigma^2_{y.x}$ is constant for all values of X, then it is called homogeneous. On the other hand, if $\sigma^2_{y.x}$ varies with X it is called heterogeneous. In Box 6.1, we present some R code to generate and plot two sets of X and Y values, one with homogenous variance and one with heterogeneous variance. In this example, the intercept (alpha) is set to 1 and the slope (beta) to 2. For the heterogeneous case, the square root of $\sigma^2_{y.x}$, i.e., $\sigma_{y.x}$, is equal to $0.5*X$. The resulting scattergrams are displayed in Fig. 6.1. Readers should experiment with this code, varying the parameters and observing the results.

Code box 6.1 R code for exploring homogeneous and heterogeneous variance for simple linear models.

```
x <- seq(from=1,to=10,by=0.02)
n <- length(x)
alpha <- 1
beta  <- 2
yhat <- alpha + beta*x
res1 <- rnorm(n,mean=0,sd=1)
y1 <- yhat+res1
res2 <- rnorm(n,mean=0,sd=1)*0.5*x
y2 <- yhat+res2
maxy <- max(c(max(y1),max(y2)))
par(mfrow=c(2,2))
plot(x,y1,ylab="Y",xlab="X",pch=".",
    main="homogeneous variance",ylim=c(0,maxy))
plot(x,y2,ylab="Y",xlab="X",pch=".",
    main="heterogeneous variance",ylim=c(0,maxy))
```

It is not unusual for scattergrams of sizes of biological organisms to display heterogeneous variance. When heterogeneity is present, it must be accounted for. Possible remedies include transforming one or both variables or modeling the conditional variance of Y as a function of X, (e.g., see Green and Valentine 1998 or Edwards and Jannink 2006).

6.1 Simple Linear Model: Trees Data

Consider the base R data set "trees." This contains observations on volume (V), girth (i.e., diameter (D)), and height (H) for 32 black cherry (*Prunus serotina*) trees. The measurement units are not contained in the data set, but they are presumably ft^3,

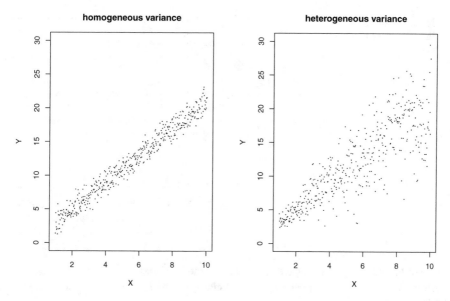

Fig. 6.1 Scattergrams of generated x and y values displaying homogeneous and heterogeneous variance

in, and *ft*, respectively. Individual tree volume is a vitally important measurement for determining the worth of a tree. It is also highly correlated with biomass which is crucially important in studies of climate change. However, volume is difficult to measure in the field. Hence, it is usual for foresters to develop models to predict the hard-to-measure V from the easy-to-measure D and the less-hard-to-measure H[1] (see, e.g., Avery et al. 2019). Cubic volume is 3-dimensional, diameter-squared is two dimensional, and height is one dimensional. Hence it is usual to multiply the latter two variables and model V as a function of the "new" variable D^2H. The R code in Box 6.2 produces the scattergram of V *versus* D^2H shown in Fig. 6.2 for the trees data.

Code box 6.2 R code plotting a scattergram of V versus D^2H for the data in the R data set trees.

```
h <- trees$Height
d <- trees$Girth
v <- trees$Volume
d2h <- d*d*h
par(mar=c(5.1,5.1,4.1,2.1))
plot(d2h,v,ylab = expression(paste("V","
  ("," ft"^"3",")")),
    xlab= expression(paste( "D"^"2","H","
      (in"^"2","ft)")))
```

---

[1] Measuring H on standing trees is not without difficulties, but it is easier than measuring V.

Fig. 6.2 Scattergram of V
versus D^2H for R data set
trees

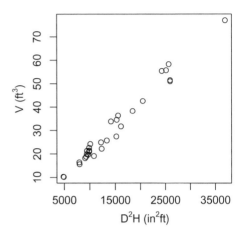

For the sake of generality, let us denote volume by Y and D^2H by X, i.e., y_i and x_i are the volume and (diameter-squared $\times$ height) of tree i. The scattergram suggests a linear relationship, i.e.,

$$y_i = \alpha + \beta x_i + e_i, \ i = 1, 2, \ldots, n, \tag{6.1.1}$$

where e_i is the unobserved random error for tree i, and n is the sample size. To complete the statistical model, we must specify a distribution for e_i. Figure 6.2 does not display heterogeneous variance, so we will proceed with the assumption that, for this data set, the conditional variance of Y is constant with respect to X. It is usual to assume that the unobserved errors follow a normal distribution with a mean of 0 and we will make that assumption here. Hence our completed sampling model is:

$$y_i = \alpha + \beta x_i + e_i, \tag{6.1.2}$$

$$e_i \sim \mathbf{N}(0, \sigma^2), \tag{6.1.3}$$

$$i = 1, 2, \ldots, n. \tag{6.1.4}$$

While models Eqs. 6.1.2–6.1.3 are correct, we prefer the following equivalent notation:

$$y_i \mid x_i, \mu_i, \sigma^2 \sim \mathbf{N}(\mu_i, \sigma^2), \tag{6.1.5}$$

$$\mu_i = \alpha + \beta x_i, \tag{6.1.6}$$

$$i = 1, 2, \ldots, n. \tag{6.1.7}$$

Our preference for the latter notation stems from our belief that the former notation lends itself to thinking of the model as a "line," whereas the latter leads to thinking of the model as expressing the conditional mean of the dependent variable. Again, each method of thinking is correct, but the latter lends itself more readily to hierarchical modeling. Also note that the parameter σ^2 in Eq. 6.1.5 is the conditional variance which we described as $\sigma^2_{y \cdot x}$ in the previous section; since the variance is homogeneous we have suppressed the subscript here for brevity.

All that remains is to specify prior distributions for the unknown parameters α, β, and σ^2. Now, foresters have been fitting volume equations to data on V, D, and H for a long time and hence if we were so inclined we *could* specify informative priors, at least for α and β. However, experience has taught us that with a sample size of $n = 32$, the data are likely to overwhelm the prior. Hence we prefer to specify non-informative priors for all 3 parameters:

$$\alpha \sim \mathbf{N}(0, 1.0 \times 10^6), \tag{6.1.8}$$

$$\beta \sim \mathbf{N}(0, 1.0 \times 10^6), \tag{6.1.9}$$

$$\sigma^2 \sim \mathbf{Ga}^{-1}(0.001, 0.001). \tag{6.1.10}$$

Recall from Chap. 2 that if $\sigma^2 \sim \mathbf{Ga}^{-1}(0.001, 0.001)$, i.e., if σ^2 follows an inverse-gamma distribution with parameters 0.001 and 0.001, then the precision, or inverse of the variance, follows a gamma distribution with parameters 0.001 and 0.001. In other words, if $\tau = \sigma^{-2}$, then $\tau \sim \mathbf{Ga}(0.001, 0.001)$. In Fig. 6.3 we display the vague $\mathbf{N}(0, 1.0 \times 10^6)$ and $\mathbf{Ga}(0.001, 0.001)$ densities. The priors for α and β are

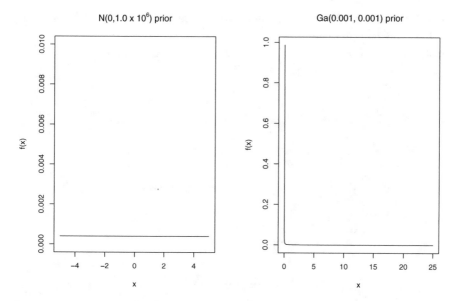

Fig. 6.3 Vague Normal and Gamma priors for trees data

essentially flat for all values, whereas the prior for τ is flat for all values except *very* small values (representing very low precision) which are unlikely to be supported by the data.

The R code to produce the plots in Fig. 6.3 is presented below in Box 6.3. We encourage readers to *always* plot their prior distributions before performing a Bayesian analysis.

Code box 6.3 R code for plotting two vague prior densities; $\mathbf{N}(0, 1.0 \times 10^{-6})$ and $\mathbf{Ga}(0.001, 0.001)$.

```
x  <- seq(from=-5,to=5,by=0.1)
var <- 1.0E6
std <- sqrt(var)
f <- dnorm(x, mean=0, sd=std, log = FALSE)
par(mar=c(5.1,5.1,4.1,2.1),mfrow=c(2,2))
plot(x,f,ylab = expression(paste("f(","x",")")),
    xlab=
        expression(paste("x")),type="l",ylim=c(0,0.01),
    main= expression(paste("N","(0,", "1.0 x
        ",10^6,") prior")))
alpha0 <- 0.001
beta0  <- 0.001
x  <- seq(from=0.001,to=25.0,by=0.001)
y  <- dgamma(x, shape = alpha0, rate = beta0, log =
    FALSE)
y_up <- max(y)
y_lo <- min(y)
plot(x,y,type="l",xlab="x",ylab="f(x)",
    xlim=c(0,25), ylim=c(y_lo,y_up),
    main=expression(paste("Ga(0.001, 0.001) prior")))
```

Equations 6.1.5–6.1.10 define the complete statistical model. The OpenBUGS code to fit the model is in Box 6.4.

Code box 6.4 OpenBUGS code for fitting simple linear model $y_i = \beta_0 + \beta_1 x_i + e_i$.

```
model
{
    ybar <- mean(y[])
    for( j in 1 : N ) {
        x[j] <- d[j]*d[j]*h[j]
    }
    for( i in 1 : N ) {
        y[i] ~ dnorm(mu[i],tau)
        mu[i] <- alpha + beta * x[i]
        num[i] <- (y[i] - mu[i])*(y[i] - mu[i])
        den[i] <- (y[i] - ybar)*(y[i] - ybar)
    }
    alpha ~ dnorm(0, 1.0E-6)     # vague priors
    beta ~ dnorm(0, 1.0E-6)
    tau ~ dgamma( 0.001, 0.001 )
    sigma <- 1 / sqrt(tau)
    rsq <- 1 -sum( num[] ) / sum( den[] )    # compute
        r-square
}
```

```
Inits
    list(tau = 1, alpha=1, beta=1)

Data
    list(N=31)
        d[]        h[]        y[]
        8.3        70        10.3
        8.6        65        10.3
        8.8        63        10.2
         .          .          .

         .          .          .

         .          .          .
       20.6        87        77.0
END
```

We do not show the full data set in the code; all 31 observations are available in R. Note also that we use two methods for inputting the data. The list format is used for the sample size (N=31) and the rectangular format is used for the actual data. If the model is fitted using the **OpenBUGS** editor, then two steps are required to input the data; one for the list portion and one for the rectangular format portion. See the **OpenBUGS** User Manual for more information on data formatting. In addition to fitting the model, the code in Box 6.1 also computes a *pseudo-r^2* value. In classical statistics,

$$r^2 = 1 - \frac{\sum_{i=1}^{n}(y_i - \hat{y}_i)^2}{\sum_{i=1}^{n}(y_i - \bar{y})^2}, \tag{6.1.11}$$

where $\hat{y}_i = \hat{\alpha} + \hat{\beta}x_i$; $\hat{\alpha}$ and $\hat{\beta}$ are the least squares estimates of α and β [2]; and $\bar{y}$ is the usual sample mean of Y. The r^2 statistic is interpreted as the percentage of variation explained by the model and is constrained to lie in the interval $(0, 1)$. To understand this, observe that $\sum(y_i - \bar{y})^2 = \sum(y_i - \hat{y}_i)^2 + \sum(\hat{y}_i - \bar{y})^2$. If we didn't know the model, our best guess for the mean of Y_i would be $\bar{y}$, and the sum of squared errors would thus be $\sum(y_i - \bar{y})^2$. However, when we have a linear model, our best guess for the mean of Y_i is $\hat{y}_i$, and the sum of squared errors would thus be $\sum(y_i - \hat{y}_i)^2$. The difference between $\sum(y_i - \bar{y})^2$ and $\sum(y_i - \hat{y}_i)^2$ is $\sum(\hat{y}_i - \bar{y})^2$, and is considered to be the reduction in error "explained" by the model.

All things considered, we prefer a higher r^2. However, there are well-known disadvantages to r^2. In particular, larger models always have r^2 values as large or larger than smaller models nested within them. Hence in a model choice setting, r^2 always chooses the larger model. However, we still find r^2 a useful general measure of how well a model fits. To that end we compute a *pseudo-r^2* on each iteration of the MCMC routine. We consider the posterior mean *pseudo-r^2* an informative measure of how well the model fits the data.

In Box 6.4, the (conditional) normal distribution on y[i] is expressed as a function of its mean and *precision*, τ. This is an **OpenBUGS** convention. However, we also compute the standard deviation (sigma). This is equal to the square root of the

---

[2]See Draper and Smith (1998) among many others for more on least squares estimation.

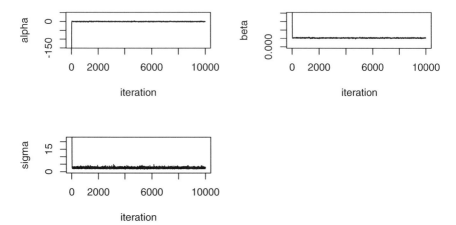

Fig. 6.4 History of alpha, beta, and sigma

Table 6.1 Posterior means, standard deviations, 0.025 quantiles and 0.975 quantiles for parameters in simple individual tree volume model

Parameter	Mean	Std dev	0.025	0.975
α	–0.2948	0.9994	–2.3020	1.6700
β	0.0021	0.0001	0.0020	0.0022
σ	2.5626	0.3497	1.9890	3.3470

variance, or equivalently, the inverse of the square root of the precision. We find it easier to understand the standard deviation than the variance or precision because the standard deviation is in the same units as the data whereas the variance is in squared units and the precision is in (squared units)$^{-1}$.

In our experience, MCMC samplers for simple models with normally distributed errors, such as the one in Box 6.4, converge very quickly. In this case, a look at the initial histories of alpha, beta, sigma, and r^2 (iterations 1:1000) in Fig. 6.4 shows that the sampler converged to the joint posterior sample almost immediately. However, computing was rapid and cheap, so we ran the sampler for 25,000 iterations and discarded the first 10,000.

The posterior distributions for α, β, and σ are shown in Fig. 6.5, and the mean, standard deviation, 0.025 quantile and 0.975 quantile of each parameter are in Table 6.1. The mean r^2 was 0.98, indicating that the model fit the data very well (not uncommon for an individual tree volume equation). As a further check on the adequacy of the model, we computed posterior predictive distributions (PPDs) (one for each of the 10,000 observations in the joint posterior sample) and compared them to the observed data. We used the R package **bayesplot** to plot a) kernel densities for 50 randomly chosen PPDs and the observed volumes; and b) boxplots for the observed volumes and 10 randomly selected PPDs. These are displayed in Fig. 6.6.

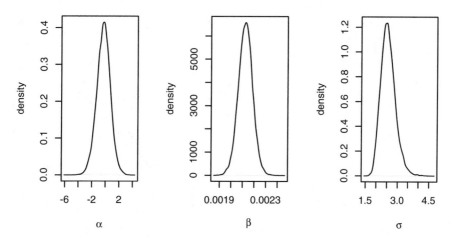

Fig. 6.5 Marginal posterior densities for α, β, and σ

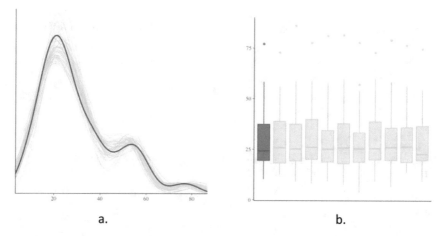

a. b.

Fig. 6.6 Kernel densities (**a**) and boxplots (**b**) for observed volumes and posterior predicted volumes using trees data

R code for producing posterior densities as in Fig. 6.5 and kernel density estimates and boxplots as in Fig. 6.6 is presented in Box 6.5.

The kernel density and boxplot from the observed volumes are in dark gray, while those for the PPDs are light gray. The kernel densities in Fig. 6.6a show that the observed data does seem to be a realization from the set of potential PPDs, as do the boxplots in Fig. 6.6b. The *pseudo-r*^{2} and the information in Fig. 6.6 demonstrate that our proposed model is a good choice for the trees data[3]. In Fig. 6.7 we show the tree data with the fitted regression line, where the latter was obtained using the posterior means of α and β.

---

[3]The posterior density for α in Fig. 6.5 indicates that a no-intercept model might also be a consideration.

Fig. 6.7 Scattergram of V
versus D^2H for R data set
trees with fitted regression

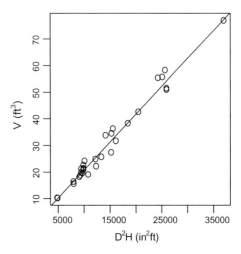

Code box 6.5   R code for producing posterior densities and kernel densities boxplots for PPDs and
observed data.

```
rm(list=ls())
#
#  Be sure to set working directory!!
#
  setwd("   ")
  v <- trees$Volume
  d <- trees$Girth
  h <- trees$Height
  d2h <- d*d*h
  Nobs <- length(d) # number of observations
library("bayesplot")
library("coda")
library("loo")
  k <- 15000  # posterior sample size
# read in posterior samples (change file name as
   appropriate)
  y <- read.table("trees.out",header=FALSE,row.names
     = NULL)
  alpha <- y[1:k,2]
  beta <- y[(k+1):(2*k),2]
  rsq <- y[((2*k)+1):(3*k),2]
  sigma <- y[((3*k)+1):(4*k),2]
  rm(y)
  k <- 15000   # k = posterior sample size
  par(mfrow=c(1,3))
  a = density(alpha)
  b = density(beta)
  c = density(sigma)
  plot(a,main = "  ",xlab=expression(alpha),
     ylab="density",lwd=1)
```

```
plot(b,main = " ",xlab=expression(beta),
    ylab="density",lwd=1)
plot(c,main = " ",xlab=expression(sigma),
    ylab="density",lwd=1)
vrep <- matrix(0,nrow=k,ncol=Nobs)
for (i in 1:k){
   mu <- alpha[i] + beta[i]*(d2h)
   vrep[i,] <- rnorm(n=Nobs,mean=mu,sd=sigma[i])
}
color_scheme_set("brightblue")
ind <- sample(1:15000, 50, replace=FALSE)
ppc_dens_overlay(v, vrep[ind, ])
ind2 <- sample(1:15000, 10, replace=FALSE)
ppc_boxplot(v, vrep[ind2, ],notch=FALSE)
```

6.1.1 Predicting a New Observation

Now that we have a fitted model, how can we use it? For example suppose a new tree has a diameter (d_{new}) of 15 inches and a height (h_{new}) of 75 feet, and we wish to predict the volume of the new tree. Call the predicted value $\hat{y}_{new}$, and let $x_{new} = d_{new}^2 h_{new}$. There are **two** sources of uncertainty to be concerned with: the uncertainty in the mean volume of the new tree (μ_{new}); and the variability of $\hat{y}_{new}$ given μ_{new}. As the sample size goes to infinity our uncertainty in μ_{new} and σ^2 goes to zero, but the variability of ($\hat{y}_{new} \mid \mu_{new}$) remains; the latter is the variability in Y unaccounted for by the model and the independent variable.

The output from an MCMC sampler such as **OpenBUGS** is conducive to generating predictive posterior distributions for new data. We simply output the joint distribution for the model parameters (α, β, σ^2) and then for each set of (α, β, σ^2) values in the joint posterior, generate a predicted value y_{pred} by sampling from a $N(\mu_{new}, \sigma^{[i]2})$ density, where $\mu_{new} = \alpha^{[i]} + \beta^{[i]} x_{new}$, $\alpha^{[i]}$, $\beta^{[i]}$, and $\sigma^{[i]2}$ are the ith values in the joint posterior sample, $i = 1, 2, \ldots, k$, and k is the posterior sample size. The generated y_{pred} values then constitute the predictive posterior distribution for volume of the new tree.

The R code in Box 6.5 is easily modified to produce predictive posterior distributions for the volume of a new tree:

Code box 6.6 R code for producing posterior predictive distributions for volume of a new tree.

```
rm(list=ls())
par(mfrow=c(1,1))
set.seed(17)
d <- 15
h <- 75
d2h <- d*d*h
k <- 15000   # posterior sample size
# read posterior samples
# set to directory where posterior samples are stored
#setwd("    ")
```

```
# read in posterior samples (change file name as
    appropriate)
y  <-  read.table("trees.out",header=FALSE,row.names =
    NULL)
a  <-  y[1:k,2]
b  <-  y[(k+1):(2*k),2]
# rsq <- y[((2*k)+1):(3*k),2]
sigma  <-  y[((3*k)+1):(4*k),2]
rm(y)
ynew <- rep(0,k)
for (i in 1:k){
   mu  <-  a[i]  +  b[i]*(d2h)
   ynew[i]  <-  rnorm(1,mean=mu,sd=sigma[i])
}
d <- density(ynew)
plot(d, ylab = "density",
     xlab= expression(paste( "Volume","
        ("," ft"^"3",")")),
     main="Posterior predictive distribution \n for
        new tree with D=15 and H=75")
stats <- matrix(0, nrow=1, ncol=5)
stats[1,1]  <-  mean(ynew)
stats[1,2]  <-  sqrt(var(ynew))
stats[1,3]  <-  quantile(ynew,0.025)
stats[1,4]  <-  quantile(ynew,0.5)
stats[1,5]  <-  quantile(ynew,0.975)
est <- data.frame(stats)
colnames(est)  <-  c("mean","std dev","0.025
    pct","median","0.975 pct")
est
```

The code in Box 6.6 resulted in the Fig. 6.8. The Figure shows the predictive posterior distribution of the volume of the new tree. The code also delivers the mean, median, standard deviation, 0.025 quantile and 0.975 quantile of the predictive posterior distribution. The mean predicted volume for a tree with a diameter of 15 inches and a height of 75 feet was 35.58 ft^3, and the standard deviation was 2.63 ft^3. The median was 35.56 ft^3 and the 0.025 and 97.5 percentiles were (30.36, 40.78). The latter is a 95% credible interval for the volume of a new tree with $d = 15$ and $h = 75$.

Fig. 6.8 Posterior predictive distribution for a new tree with $D = 15$ and $H = 75$

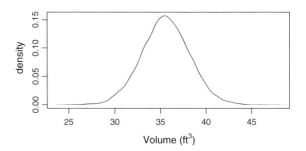

6.2 Hierarchical Linear Models

In many studies, the data occur in groups. For instance, in a field trial of a fertilizer applied to a particular plant, we may have groups of plants exposed to different levels of the fertilizer, and a control group not exposed to any. As another example, in a study of the growth rate of a particular animal, we may study animal populations in a number different habitats.

Suppose we are in such a situation and we have a dependent variable and an independent variable measured on a sample of observations from each group. We immediately see that we have at least two options: (1) we might ignore the group labels and fit one linear model to all the data, or (2) we might independently fit a separate linear model to the data from each group. However, we might not be satisfied with slavish adherence to either of these two options. We might reasonably prefer a "middle ground." If so, then a hierarchical structure provides the solution.

As mentioned in Sect. 4.4, hierarchical models are appropriate when it is reasonable to believe that the data are exchangeable (see Sect. 4.4.2). It is up to the investigator to decide whether exchangeability holds. For instance, in the case of growth rates of an animal, suppose all but one of the habitats were heavily forested, while one was moderately developed. Exchangeability might be questionable in that case. Of course one could always model the data from the questionable habitat by itself and model the other habitats as exchangeable. In general, ignorance makes exchangeability more palatable; the more we know about the groups from which data arise, the less we are usually willing to postulate exchangeability.

6.2.1 Rat Growth Data

The R package **nlme** contains the data set BodyWeight. This data set consists of weight measurements (*grams*) over time on groups of rats fed one of three diets. All rats were measured on day 1 and every 7 days thereafter until day 64, with an extra measurement at day 44. This resulted in 11 weight measurements per rat. Eight rats were fed diet 1, while 4 were fed diet 2 and 4 were fed diet 3.

6.2.2 Diet 2 Rats

In this example, we restrict our attention to the rats fed diet 2. We will consider the full data set in Sect. 6.2.4.

We have no reason to suspect that the rats are different from each other in any meaningful way, so exchangeability seems a plausible assumption for this example. A plot of the data for each rat (Fig. 6.9) suggests that a linear model is appropriate. The weights for rat 3 on days 43 and 44 appear suspect, but we shall proceed *as if*

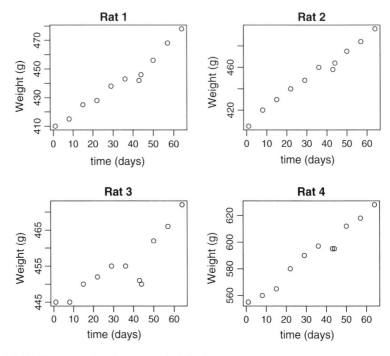

Fig. 6.9 Weight versus time for four rats fed diet 2

these data points were inspected and found to be valid. For the sake of generality, let y_{ij} be the weight (g) of rat i at time j and let x_{ij} be the time (in days) of measurement j on rat i. We will specify the following data, or sampling, model:

$$y_{ij} = \alpha_i + \beta_i x_{ij} + e_{ij}; \quad i = 1, 2, \ldots, k; \quad j = 1, 2, \ldots, n_i. \tag{6.2.1}$$

In model Eq. 6.2.1, α_i and β_i are the intercept and slope for rat i, respectively, k is the number of rats (4 in this example), n_i is the number of observations on rat i (11 for each rat in this example), and e_{ij} is the unobserved error for rat i, observation j.

To complete the data model, we need to postulate an error distribution for the e_{ij}'s. As usual, we will assume normally distributed errors, i.e., $e_{ij} \sim \mathbf{N}(0, \sigma^2)$. Again, we prefer to write

$$(y_{ij} \mid x_{ij}, \mu_{ij}, \sigma^2) \sim \mathbf{N}(\mu_{ij}, \sigma^2), \tag{6.2.2}$$

$$\mu_{ij} = \alpha_i + \beta_i x_{ij}; \quad i = 1, 2, \ldots, k; \quad j = 1, 2, \ldots, n_i. \tag{6.2.3}$$

In Eq. 6.2.2, μ_{ij} is the mean weight for rat i at measurement j. We specify a constant variance (σ^2) among rats in Eq. 6.2.2. In work not shown, we modified this to assume rat-specific variances. Examination of DIC indicated that the constant variance model was superior for this data and hence we use it here. Readers are encouraged to investigate this themselves.

Next we assume exchangeabilty and model the α_i's and β_i's as realizations from populations of intercepts and slopes, respectively:

$$(\alpha_i \mid \mu_\alpha, \sigma_\alpha^2) \sim N(\mu_\alpha, \sigma_\alpha^2), \ (\beta_i \mid \mu_\beta, \sigma_\beta^2) \sim N(\mu_\beta, \sigma_\beta^2), \ i = 1, 2, \ldots, k. \quad (6.2.4)$$

In Eq. 6.2.4, we specify independent priors for α_i and β_i. Alternatively, we could have specified a bivariate normal distribution as in Gelfand et al. (1990), and specified a prior covariance for α_i and β_i. We chose the specification in Eq. 6.2.4 for simplicity. The parameters α_i and β_i will still exhibit a covariance in the joint posterior distribution, but this will come solely from the data.

To complete our hierarchical statistical model, we need a prior on σ^2 and hyperpriors on μ_α, μ_β, σ_α^2, and σ_β^2. We will choose vague [hyper]priors for these.[4]

Following Gelfand et al. (1990) and many others, we specify a vague inverse gamma prior on σ^2, i.e.,

$$\sigma^2 \sim Ga^{-1}(0.001, 0.001). \quad (6.2.5)$$

Recall that Eq. 6.2.5 implies a $Ga(0.001, 0.001)$ prior on the precision. This is important because OpenBUGS parameterizes the normal density in terms of its mean and precision. For μ_α and μ_β we specify vague normal priors:

$$\mu_\alpha \sim N(0.0, 1.0 \times 10^6), \ \mu_\beta \sim N(0.0, 1.0 \times 10^6). \quad (6.2.6)$$

The $Ga(0.001, 0.001)$ and $N(0.0, 1.0 \times 10^6)$ densities are displayed in Fig. 6.3.

As mentioned in Sect. 4.4.2, Gelman (2006) showed that inverse-gamma priors can be problematic for variances at higher levels than the data model. Accordingly, we follow Gelman's recommendation and specify vague uniform priors for σ_α and σ_β, i.e., the standard deviations of the densities in Eq. 6.2.4:

$$\sigma_\alpha \sim Unif(0, 100), \ \sigma_\beta \sim Unif(0, 100). \quad (6.2.7)$$

Bringing the data model, priors and hyperpriors all together:
Data model:

$$(y_{ij} \mid x_{ij}, \mu_{ij}, \sigma^2) \sim N(\mu_{ij}, \sigma^2), \ \mu_{ij} = \alpha_i + \beta_i x_{ij}, \quad (6.2.8)$$

Priors:

$$(\alpha_i \mid \mu_\alpha, \sigma_\alpha^2) \sim N(\mu_\alpha, \sigma_\alpha^2), \ (\beta_i \mid \mu_\beta, \sigma_\beta^2) \sim N(\mu_\beta, \sigma_\beta^2), \quad (6.2.9)$$

$$\sigma^2 \sim Ga^{-1}(0.001, 0.001), \quad (6.2.10)$$

---

[4]As a general rule, we find it easier to specify informative priors on means and/or model coefficients in the data model than on variances or hyperparameters.

Hyperpriors:

$$\mu_\alpha \sim \mathbf{N}(0.0, 1.0 \times 10^6), \ \ \mu_\beta \sim \mathbf{N}(0.0, 1.0 \times 10^6), \quad\quad (6.2.11)$$

$$\sigma_\alpha \sim \mathbf{Unif}(0, 100), \ \ \sigma_\beta \sim \mathbf{Unif}(0, 100), \quad\quad (6.2.12)$$

$i = 1, 2, \ldots, k; \ j = 1, 2, \ldots, n_i$.

In Sect. 6.1, we introduced r^2 as a measure of the percent of variation "explained" by the model, i.e., 1 - $\big($(sum of squared errors with model) / (sum of squared errors without model)$\big)$. Here we use the same idea, with a minor twist. We must specify what would we do if we didn't have the independent variable in the model available to us. In that case, we might use the following alternative hierarchical model:

$$\textbf{Alt. Data model:} \ (y_{ij} \mid \theta_i, \sigma^2) \sim \mathbf{N}(\theta_i, \sigma^2), \quad\quad (6.2.13)$$

$$\textbf{Alt. Priors:} \ (\theta_i \mid \mu_\theta, \sigma_\theta^2) \sim \mathbf{N}(\mu_\theta, \sigma_\theta^2), \ \ \sigma^2 \sim \mathbf{Ga}^{-1}(0.001, 0.001), \quad (6.2.14)$$

$$\textbf{Alt. Hyperpriors:} \ \mu_\theta \sim \mathbf{N}(0.0, 1.0 \times 10^6), \ \ \sigma_\theta \sim \mathbf{Unif}(0, 100), \quad (6.2.15)$$

$i = 1, 2, \ldots, k; \ j = 1, 2, \ldots, n_i$.

Hence we need a measure of how our error is reduced by using μ_{ij} as the mean of Y_{ij} instead of θ_i. We thus define a *pseudo-r^2* as

$$pseudo - r^2 = 1 - \frac{\displaystyle\sum_{i=1}^{k}\sum_{j=1}^{n_i}(y_{ij} - \hat{\mu}_{ij})^2}{\displaystyle\sum_{i=1}^{k}\sum_{j=1}^{n_i}(y_{ij} - \hat{\theta}_i)^2}. \quad\quad (6.2.16)$$

In Eq. 6.2.16, $\hat{\mu}_{ij}$ will be the current value for μ_{ij} and $\hat{\theta}_i$ will be the current value for θ_i on each iteration of the MCMC sampler.

The OpenBUGS code to fit the hierarchical linear model to the data from the rats on diet 2 is presented in Box 6.7. In order to compute our *pseudo-r^2* value, we need to fit the alternative hierarchical model in which $(y_{ij} \mid \theta_i, \sigma^2) \sim \mathbf{N}(\theta_i, \sigma^2)$. To do this in OpenBUGS we input the data twice, assigning it to both the variables Y and Z. Note that we truncated the priors for the β_i's, and the hyperprior for μ_β to be positive (accomplished by adding T(0,) to the distribution declarations). We do this since we are sure that the slopes must be ≥ 0.

Code box 6.7 OpenBUGS code for fitting hierarchical linear model to data for rats fed diet 2.

```
MODEL
{
for( i in 1 : K ) {
    for( j in 1 : N ) {
    Y[i , j] ~ dnorm(mu[i , j], tau)
    mu[i , j] <- alpha[i] + beta[i] * x[j]
    Z[i , j] ~ dnorm(theta[i], prec)
    }
}
for (i in 1:K) {
    alpha[i] ~ dnorm(alpha.mean, alpha.tau)
    beta[i] ~ dnorm(beta.mean, beta.tau)T(0,)
    theta[i] ~ dnorm(theta.mean, theta.prec)
}

for (i in 1:K) {
    for (j in 1 : N) {
        numerator[i,j] <- (Y[i, j] - mu[i, j])*(Y[i,
            j] - mu[i, j])
        denominator[i,j] <- (Y[i, j] - theta[i])*(Y[i,
            j] - theta[i])
    }
}
rsq <- 1 - ( sum(numerator[ , ])  / sum(denominator[ ,
    ]) )

tau ~ dgamma(0.001,0.001)
sigma <- 1 / sqrt(tau)
alpha.mean ~ dnorm(0.0,1.0E-6)
alpha.sigma ~ dunif(1.0E-6, 500)
alpha.tau <- 1/(alpha.sigma * alpha.sigma)
beta.mean ~ dnorm(0.0,1.0E-6)T(0,)
beta.sigma ~ dunif(1,100)
beta.tau <- 1/(beta.sigma * beta.sigma)
theta.mean ~ dnorm(0.0, 1.0E-6)
prec ~ dgamma(0.001, 0.001)
theta.sig ~ dunif(1,100)
theta.prec <- 1/(theta.sig * theta.sig)
}
INITS
list(  alpha = c(1,1,1,1),
        beta  = c(1,1,1,1),
        tau = 1,
        theta=c(1,1,1,1), theta.mean=1,
        theta.sig = 1, prec = 1,
        alpha.mean = 1, beta.mean = 1,
        alpha.sigma = 1, beta.sigma = 1 )

DATA
list(
x = c(1,   8, 15, 22, 29, 36, 43, 44, 50, 57, 64),
    K=4, N=11,
```

```
Y = structure(
        .Data=c(410,   415,   425,   428,   438,   443,   442,
          446,   456,   468,   478,
                        405,   420,   430,   440,   448,   460,
                          458,   464,   475,   484,   496,
                        445,   445,   450,   452,   455,   455,
                          451,   450,   462,   466,   472,
                        555,   560,   565,   580,   590,   597,
                          595,   595,   612,   618,   628),
        .Dim=c(4,11)),
Z = structure(
        .Data=c(410,   415,   425,   428,   438,   443,   442,
          446,   456,   468,   478,
                        405,   420,   430,   440,   448,   460,
                          458,   464,   475,   484,   496,
                        445,   445,   450,   452,   455,   455,
                          451,   450,   462,   466,   472,
                        555,   560,   565,   580,   590,   597,
                          595,   595,   612,   618,   628),
        .Dim=c(4,11))
)
```

As in the example in Sect. 6.1, the MCMC sampler converged quickly. History and quantile plots in OpenBUGS indicated convergence for all parameters was obtained with 2,000 iterations. Computation time was cheap however, so we ran the sampler in Box 6.7 for 50,000 iterations and discarded the initial 25,000. The marginal posterior distributions for α_1, α_2, α_3, and α_4 are presented in Fig. 6.10, for β_1, β_2, β_3, and β_4 in Fig. 6.11, and for μ_α, μ_β, σ_α, σ_β and σ in Fig. 6.12. The posterior means, standard deviations, 0.025 and 0.975 quantiles are presented in Table 6.2. The mean *pseudo-r²* was 0.96, indicating a good fit to the data.

As a further check on the model, we generated 25,000 posterior predictive distributions. The code to do this is presented in Box 6.8. Here we had a choice: we *could* have done this rat-by-rat, i.e., generated 25,000 posterior predictive distributions for each rat. For convenience, we chose to generate 25,000 posterior predictive distributions for the entire data set. For each set of $(\alpha_i, \beta_i, \sigma^2)$ in the joint posterior sample, we generated a weight for rat i, $i = 1, 2, 3, 4$, for each the 11 days on which weights were recorded in the original data set. In this way we generated 25,000 replicate data sets. Kernel densities and boxplots of all the generated weights were plotted (using the R package **bayesplot**) along with a kernel density of the all the actual recorded weights. These are shown in Fig. 6.13. In the Fig. we show the densities and boxplots for 50 and 10 randomly chosen replicate data sets and boxplots, respectively. The density and boxplot for the observed weights are in dark gray; those for the replicate data sets are in light gray. Figure 6.13 indicates that the model is appropriate for these data.

Table 6.2 Posterior means, standard deviations, 0.025 quantiles, and 0.975 quantiles for parameters and hyperparameters in diet 2 rat growth model

[Hyper]parameter	Mean	Std dev	0.025	0.975
α_1	406.90	2.65	401.70	412.10
α_2	407.81	2.63	402.50	412.90
α_3	442.58	2.62	437.40	447.80
α_4	551.89	2.67	546.50	557.10
β_1	1.01	0.07	0.88	1.15
β_2	1.34	0.07	1.21	1.47
β_3	0.36	0.07	0.23	0.50
β_4	1.15	0.07	1.01	1.29
μ_α	449.99	73.89	292.00	599.50
μ_β	0.89	1.28	0.03	2.95
σ_α	124.30	79.51	43.36	362.00
σ_β	2.21	2.88	1.02	7.26
σ	4.35	0.53	3.45	5.52

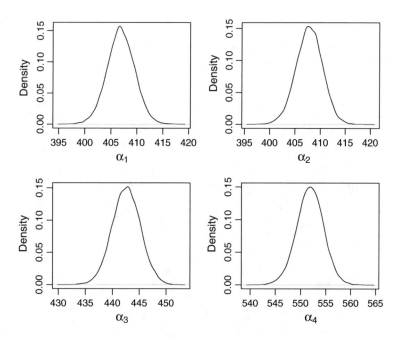

Fig. 6.10 Marginal posterior distributions for α_1, α_2, α_3, and α_4 in hierarchical linear model for rats fed diet 2

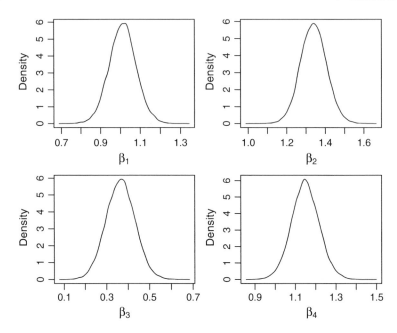

Fig. 6.11 Marginal posterior distributions for β_1, β_2, β_3, and β_4 in hierarchical linear model for rats fed diet 2

Code box 6.8 R code for generating posterior predictive distributions for rats fed diet 2.

```
rm(list=ls())
#
#   Be sure to set working directory!!
#
setwd("   ")
library(nlme)
library("bayesplot")
head(BodyWeight)
D <- BodyWeight[BodyWeight[,4]==2,]
y <- D$weight
x <- D$Time
Nobs <- length(y)
Nrats <- 4
Nperrat <- Nobs/Nrats
k <- 25000   # k = posterior sample size
alpha <- matrix(0,nrow=k,ncol=Nrats)
beta <- matrix(0,nrow=k,ncol=Nrats)
rsq <- matrix(0,nrow=k,ncol=1)
sigma <- matrix(0,nrow=k,ncol=1)
alpha_mu <- rep(0,k)
alpha_sig <- rep(0,k)
beta_mu <- rep(0,k)
beta_sig <- rep(0,k)
```

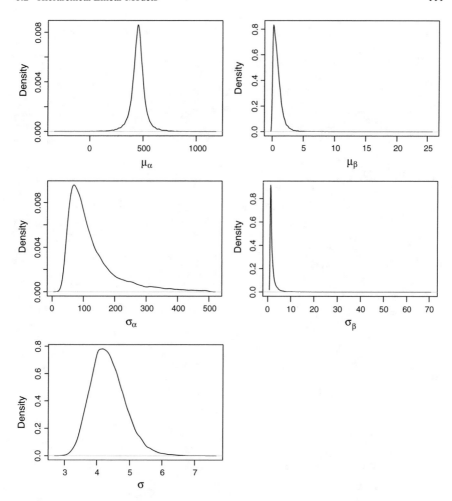

Fig. 6.12 Marginal posterior distributions for μ_α, μ_β, σ_α, σ_β, and σ in hierarchical linear model for rats fed diet 2

```
# read posterior samples

z <- read.table("rat diet
    2.out",header=FALSE,row.names = NULL)
alpha[,1] <- z[1:k,2]
alpha[,2] <- z[(k+1):(2*k),2]
alpha[,3] <- z[((2*k)+1):(3*k),2]
alpha[,4] <- z[((3*k)+1):(4*k),2]
alpha_mu[] <- z[((4*k)+1):(5*k),2]
alpha_sig[] <- z[((5*k)+1):(6*k),2]
beta[,1] <- z[((6*k)+1):(7*k),2]
beta[,2] <- z[((7*k)+1):(8*k),2]
beta[,3] <- z[((8*k)+1):(9*k),2]
beta[,4] <- z[((9*k)+1):(10*k),2]
beta_mu[] <- z[((10*k)+1):(11*k),2]
```

```
beta_sig[] <- z[((11*k)+1):(12*k),2]
rsq <- z[((12*k)+1):(13*k),2]
sigma <- z[((13*k)+1):(14*k),2]
rm(z)

yrep1 <- matrix(0,nrow=k,ncol=Nperrat)
yrep2 <- matrix(0,nrow=k,ncol=Nperrat)
yrep3 <- matrix(0,nrow=k,ncol=Nperrat)
yrep4 <- matrix(0,nrow=k,ncol=Nperrat)


for (i in 1:k){
  mu1 <- alpha[i,1] + beta[i,1]*x[1:11]
  yrep1[i,] <- rnorm(n=Nperrat,mean=mu1,sd=sigma[i])
  mu2 <- alpha[i,2] + beta[i,2]*x[12:22]
  yrep2[i,] <- rnorm(n=Nperrat,mean=mu2,sd=sigma[i])
  mu3 <- alpha[i,3] + beta[i,3]*x[23:33]
  yrep3[i,] <- rnorm(n=Nperrat,mean=mu3,sd=sigma[i])
  mu4 <- alpha[i,4] + beta[i,4]*x[34:44]
  yrep4[i,] <- rnorm(n=Nperrat,mean=mu4,sd=sigma[i])
}
wrep <- cbind(wrep1,wrep2,wrep3,wrep4)

color_scheme_set("brightblue")
ind <- sample(1:k, 50, replace=FALSE)
ppc_dens_overlay(y, yrep[ind, ])
ind2 <- sample(1:k, 10, replace=FALSE)
ppc_boxplot(y, yrep[ind2, ],notch=FALSE)
```

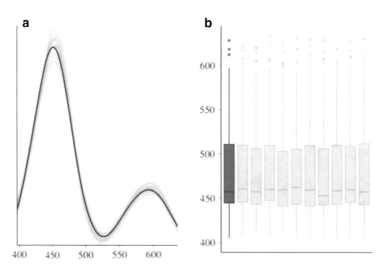

Fig. 6.13 Kernel densities (a) and boxplots (b) for observed weights and posterior predicted weights using data from rats fed diet 2

6.2.3 Predicting a New Observation

Now that we have fitted the model, we would presumably want to use it. Two possible situations immediately arise: i) We may want to predict a future weight for one of the rats included in the data, or ii) We may want to predict the weight at a specified time for a new, randomly chosen rat. Since we are not likely to have particular interest in any of the 4 rats represented in the data, ii) is much more likely than i), but we will include both for completeness. In the terminology of Sect. 6.2, rats in this example are the "groups," i.e., we have data on each of 4 groups (rats).

6.2.3.1 Predicting a New Observation for a Group Observed in the Data

Suppose we wish to predict the weight of rat 1 on day 70. As in Sect. 6.1.1, we have two sources of variability to be concerned with. First, we must account for the variability of α_1 and β_1; and second, we must account for the variability of $y_{1,new}$ given $(x_{new} = 70, \alpha_1, \beta_1, \sigma^2)$. For brevity, we will refer to $y_{1,new}$ as y_{new} for the remainder of this section.

Given the output of our MCMC sampler, we account for the variability in α_1 and β_1 by computing $\mu_{new}^{[i]} = \alpha_1^{[i]} + \beta_1^{[i]} x_{new}$ for each $(\alpha_1^{[i]}, \beta_1^{[i]})$ pair in the joint posterior sample. Then we account for the variability in $(y_{new} \mid x_{new}, \alpha_1, \beta_1, \sigma^2)$ by generating a value for y_{new} from a $\mathbf{N}(\mu_{new}^{[i]}, \sigma^{[i]2})$ distribution, where $\sigma^{[i]2}$ is the ith value of σ^2 in the joint posterior sample. The resulting values are a sample from the posterior predictive distribution for the weight of that particular rat on the specified day (70). The R code to implement this is in Box 6.9, and the resulting predictive distribution is shown in Fig. 6.14. The mean predicted weight of rat 1 at age 70 is 477.6 g, with a standard deviation of 5.2. The 95% credible interval is (467.3, 487.8), formed by the 0.025 and 0.975 quantiles of the sample from the predictive distribution.

Code box 6.9 R code for generating posterior predictive distributions for rat 1 at age 70.

```
rm(list=ls())
#
#   Be sure to set working directory!!
#
setwd("   ")

k <- 25000   # k = posterior sample size
alpha <- matrix(0,nrow=k,ncol=1)
beta  <- matrix(0,nrow=k,ncol=1)
sigma <- matrix(0,nrow=k,ncol=1)

# read posterior samples

y <- read.table("rat diet
    2.out",header=FALSE,row.names = NULL)
alpha[] <- y[1:k,2]
beta[]  <- y[((6*k)+1):(7*k),2]
sigma <- y[((13*k)+1):(14*k),2]
```

```
rm(y)
yhat <- rep(0, nrow = k, ncol = 1)

x <- 70
muhat <- alpha[,1] + beta[,1]*x
yhat <- rnorm(n=k,mean=muhat[],sd=sigma[])
stats <- matrix(0,nrow=1,ncol=4)
stats[1,1] <- mean(yhat)
stats[1,2] <- sqrt(var(yhat))
stats[1,3] <- quantile(yhat,0.025)
stats[1,4] <- quantile(yhat,0.975)
est <- data.frame(stats)
colnames(est) <- c("mean","std dev","0.025
    pct","0.975 pct")
est
d <- density(yhat)
plot(d, ylab = "density",
     xlab= "predicted weight at age 70 for rat
        1",main=" ")
```

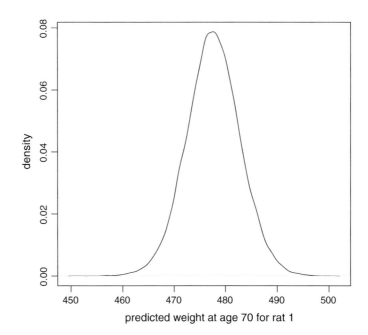

Fig. 6.14 Posterior predictive distribution for weight of rat 1 at age 70

6.2.3.2 Predicting a New Observation for a Group Not Observed in the Data

Suppose we wanted to predict the weight at day 70 of a new, independent rat fed diet 2. For this new rat, the posterior distributions of α_i and $\beta_i,$, $i = 1, 2, 3, 4$ are irrelevant. Instead we must generate a value for α from the joint posterior sample of μ_α and σ_α, and a value for β from the joint posterior sample of μ_β and σ_β. This is because in our model we postulated that the α's are realizations from a $\mathbf{N}(\mu_\alpha, \sigma_\alpha)$ density and the β's are realizations from a $\mathbf{N}(\mu_\beta, \sigma_\beta)$ density.

Let the superscript "[j]" indicate the jth value in the posterior sample, $j = 1, 2, \ldots, k$. For each pair of $\mu_\alpha^{[j]}$ and $\sigma_\alpha^{[j]}$ in our joint posterior sample, we generate an α from a $\mathbf{N}(\mu_\alpha^{[j]}, \sigma_\alpha^{[j]})$ density. Similarly, for each pair of $\mu_\beta^{[j]}$ and $\sigma_\beta^{[j]}$ in our joint posterior sample, we generate a β from a $\mathbf{N}(\mu_\beta^{[j]}, \sigma_\beta^{[j]})$ density. Call the generated α and β values α^* and β^*, respectively. Finally, we generate a predicted value of y from a $\mathbf{N}(\mu^*, \sigma^{2[j]})$ distribution where $\mu^* = \alpha^* + \beta^* \times 70$ and $\sigma^{2[j]}$ is the jth value of σ^2 in the posterior sample, $j = 1, 2, \ldots, k$.

The joint posterior samples of $(\mu_\alpha, \sigma_\alpha, \mu_\beta, \sigma_\beta)$ are based on all the observed data, i.e., data from all four rats. As a consequence, we expect the 95% credible intervals to be wider than the corresponding intervals for one of the observed rats.

In Box 6.10 we present the R code to generate a sample from the predictive posterior distribution of a randomly chosen, independent rat. The resulting predictive density is in Fig. 6.15. The mean predicted weight of an independent, randomly

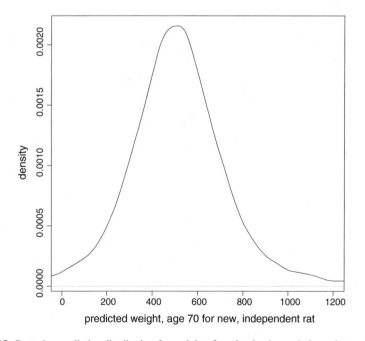

Fig. 6.15 Posterior predictive distribution for weight of randomly chosen independent rat at age 70

chosen rat at age 70 is 510.9 g, with a standard deviation of 265.6 The 95% credible interval is (24.8, 1047.0), formed by the 0.025 and 0.975 quantiles of the sample from the predictive distribution. There is considerably more uncertainty regarding the weight of an independently chosen rat than there is of the weight of a rat observed in the sample. This is due to the uncertainty about $(\mu_\alpha, \sigma_\alpha, \mu_\beta, \sigma_\beta)$, as reflected in Fig. 6.12 and Table 6.2.

Code box 6.10 R code for generating posterior predictive distributions for an independent, randomly chosen rat at age 70.

```
rm(list=ls())
set.seed(17)
#
#  Be sure to set working directory!!
#
setwd("  ")

k <- 25000  # k = posterior sample size
alpha_mu <- matrix(0,nrow=k,ncol=1)
alpha_sig <- matrix(0,nrow=k,ncol=1)
beta_mu <- matrix(0,nrow=k,ncol=1)
beta_sig <- matrix(0,nrow=k,ncol=1)
sigma <- matrix(0,nrow=k,ncol=1)
muhat <- matrix(0,nrow=k,ncol=1)
# read posterior samples
y <- read.table("rat diet
    2.out",header=FALSE,row.names = NULL)
alpha_mu[]  <- y[((4*k)+1):(5*k),2]
alpha_sig[]  <- y[((5*k)+1):(6*k),2]
beta_mu[]  <- y[((10*k)+1):(11*k),2]
beta_sig[]  <- y[((11*k)+1):(12*k),2]
sigma <- y[((13*k)+1):(14*k),2]
rm(y)
yhat <- rep(0, nrow = k, ncol = 1)
x <- 70
for (i in 1:k){
alpha <- rnorm(n=1,mean=alpha_mu[i],sd=alpha_sig[i])
beta <- rnorm(n=1,mean=beta_mu[i],sd=beta_sig[i])
muhat[i]  <- alpha + beta*x
yhat[i]  <- rnorm(n=i,mean=muhat[i],sd=sigma[i])
}
stats <- matrix(0,nrow=1,ncol=4)
stats[1,1]  <- mean(yhat)
stats[1,2]  <- sqrt(var(yhat))
stats[1,3]  <- quantile(yhat,0.025)
stats[1,4]  <- quantile(yhat,0.975)

est <- data.frame(stats)
colnames(est)  <- c("mean","std dev","0.025
    pct","0.975 pct")
est

d <- density(yhat,from=quantile(yhat,0.01),
             to=quantile(yhat,0.99),kernel="optcosine")
```

```
plot(d, ylab = "density",
     xlab= "predicted weight, age 70 for new,
        independent rat",main=" ",xlim=c(0,1200))
```

6.2.4 Full Rat Data Set

In this section, we expand the model of Sect. 6.2.1 to include all the data, i.e., the data on the 8 rats fed diet 1 and the 4 rats on diet 3 as well as the 4 on diet 2.

In our data model, we model the weight for rat j, fed diet i at measurement time k as follows:

Data model:

$$(y_{ijk} \mid x_{ijk}, \mu_{ijk}, \sigma_i^2) \sim \mathbf{N}(\mu_{ijk}, \sigma_i^2), \tag{6.2.17}$$

$$\mu_{ijk} = \alpha_{ij} + \beta_{ij} x_{ijk}, \tag{6.2.18}$$

where x_{ijk} = time (days) of measurement k on rat j and diet i; $i = 1, 2, 3$; $j = 1, 2, \ldots, n_i$; and $k = 1, 2, \ldots, n_{ij}$. The number of rats fed diet i is n_i and the total number of measurements on rat j, diet i is n_{ij}. In Eq. 6.2.17 we specify a homogeneous variance for all the rats fed each diet, i.e., a diet-specific variance. We also investigated a constant variance for all measurements, and rat-specific variance for each diet. DIC favored Eq. 6.2.17 so we use it here. Based on plots of the raw data, we did not feel a heterogeneous variance (a variance that changed with x_{ijk}) was warranted.

The data model (6.2.17) includes a separate intercept and slope for each diet-rat combination. In our prior distributions, we will model the intercepts and slopes for the rats for each diet as realizations from normal densities with means and variances specific to each diet:

Prior distributions:

$$(\alpha_{ij} \mid \mu_{\alpha i}, \sigma_{\alpha i}^2) \sim \mathbf{N}(\mu_{\alpha i}, \sigma_{\alpha i}^2), \tag{6.2.19}$$

$$(\beta_{ij} \mid \mu_{\beta i}, \sigma_{\beta i}^2) \sim \mathbf{N}(\mu_{\beta i}, \sigma_{\beta i}^2). \tag{6.2.20}$$

In Eqs. 6.2.19 and 6.2.20, $\mu_{\alpha i}$ and $\mu_{\beta i}$ are the mean intercept and slope for rats fed diet i, respectively. If the aim of the experiment was to determine if rats grew differently under the 3 diets, then the $\mu_{\alpha i}$'s and $\mu_{\beta i}$'s would be the parameters of ultimate interest.

As in the previous section, we will assume a vague inverse gamma prior distribution on the variance of the data model (σ_i^2):

$$\sigma_i^2 \sim \mathbf{Ga}^{-1}(0.001, 0.001). \tag{6.2.21}$$

For hyperpriors, we will assume that $\mu_{\alpha i}$ and $\mu_{\beta i}$ are themselves generated from normal distributions:

$$(\mu_{\alpha i} \mid \mu_\alpha, \sigma_\alpha^2) \sim \mathbf{N}(\mu_\alpha, \sigma_\alpha^2), \quad (\mu_{\beta i} \mid \mu_\beta, \sigma_\beta^2) \sim \mathbf{N}(\mu_\beta, \sigma_\beta^2). \tag{6.2.22}$$

As before we specify vague uniform densities on $\sigma_{\alpha i}$ and $\sigma_{\beta i}$:

$$\sigma_{\alpha i} \sim \mathbf{Unif}(0, 500), \quad \sigma_{\beta i} \sim \mathbf{Unif}(0, 500). \tag{6.2.23}$$

Finally, we specify vague normal hyper-hyperpriors on μ_α and μ_β, and vague uniform hyper-hyperpriors on σ_α and σ_β:

$$\mu_\alpha \sim \mathbf{N}(0, 1.0 \times 10^6), \quad \mu_\beta \sim \mathbf{N}(0, 1.0 \times 10^6), \tag{6.2.24}$$

$$\sigma_\alpha \sim \mathbf{Unif}(0, 1000), \quad \sigma_\beta \sim \mathbf{Unif}(0, 1000). \tag{6.2.25}$$

The limits on the Uniform densities in Eqs. 6.2.23 and 6.2.25 were chosen after experimentation revealed that lower bounds on these distributions caused the joint posterior distributions to be truncated at the bounds.

Bringing the data model, priors, hyperpriors and hyper-hyperpriors all together:

Data model:

$$(y_{ijk} \mid x_{ijk}, \mu_{ijk}, \sigma_i^2) \sim \mathbf{N}(\mu_{ijk}, \sigma_i^2), \tag{6.2.26}$$

$$\mu_{ijk} = \alpha_{ij} + \beta_{ij} x_{ijk}. \tag{6.2.27}$$

Prior distributions:

$$(\alpha_{ij} \mid \mu_{\alpha i}, \sigma_{\alpha i}^2) \sim \mathbf{N}(\mu_{\alpha i}, \sigma_{\alpha i}^2), \quad (\beta_{ij} \mid \mu_{\beta i}, \sigma_{\beta i}^2) \sim \mathbf{N}(\mu_{\beta i}, \sigma_{\beta i}^2), \tag{6.2.28}$$

$$\sigma_i^2 \sim \mathbf{Ga}^{-1}(0.001, 0.001). \tag{6.2.29}$$

Hyperprior distributions:

$$(\mu_{\alpha i} \mid \mu_\alpha, \sigma_\alpha^2) \sim \mathbf{N}(\mu_\alpha, \sigma_\alpha^2), \quad (\mu_{\beta i} \mid \mu_\beta, \sigma_\beta^2) \sim \mathbf{N}(\mu_\beta, \sigma_\beta^2), \tag{6.2.30}$$

$$\sigma_{\alpha i} \sim \mathbf{Unif}(0, 500), \quad \sigma_{\beta i} \sim \mathbf{Unif}(0, 500). \tag{6.2.31}$$

Hyper-hyperprior distributions:

$$\mu_\alpha \sim \mathbf{N}(0, 1.0 \times 10^6), \quad \mu_\beta \sim \mathbf{N}(0, 1.0 \times 10^6), \tag{6.2.32}$$

$$\sigma_\alpha \sim \mathbf{Unif}(0, 1000), \quad \sigma_\beta \sim \mathbf{Unif}(0, 1000). \tag{6.2.33}$$

Model Eq. 6.2.26 along with the priors, hyperpriors and hyper-priors can be implemented in OpenBUGS with the code in Box 6.11.

Code box 6.11 OpenBUGS code for fitting hierarchical linear model to complete rat data set.

```
MODEL
# constant variance per diet
####################################
# sampling model
{
for( i in 1 : N1 ) {
    y[ i ] ~ dnorm(mu[i], tau[ diet[i] ])
    mu[ i ] <- alpha1[rat[i]] + ( beta1[rat[i]] * x[i]
        )
}

for( i in (N1+1) : (N1+N2) ) {
    y[ i ] ~ dnorm(mu[i], tau[ diet[i] ])
    mu[ i ] <- alpha2[rat[i]] + ( beta2[rat[i]] * x[i]
        )
}

for( i in (N1+N2+1) : (N1+N2+N3) ) {
    y[ i ] ~ dnorm(mu[i], tau[ diet[i] ])
    mu[ i ] <- alpha3[rat[i]] + ( beta3[rat[i]] * x[i]
        )
}

####################################
# priors
for (j in 1:8) {
    alpha1[j] ~ dnorm(alpha.mean[1], alpha.tau[1])
    beta1[j] ~ dnorm(beta.mean[1], beta.tau[1])
#   beta1[j] ~ dnorm(beta.mean[1], beta.tau[1])T(0,)
}

for (j in 1:4){
    alpha2[j] ~ dnorm(alpha.mean[2], alpha.tau[2])
    beta2[j] ~ dnorm(beta.mean[2], beta.tau[2])
#   beta2[j] ~ dnorm(beta.mean[2], beta.tau[2])T(0,)
    alpha3[j] ~ dnorm(alpha.mean[3], alpha.tau[3])
    beta3[j] ~ dnorm(beta.mean[3], beta.tau[3])
#   beta3[j] ~ dnorm(beta.mean[3], beta.tau[3])T(0,)
}
for (i in 1:ndiet) {
    tau[i] ~ dgamma(0.001,0.001)
    sigma[i] <- 1 / sqrt(tau[i])
}

####################################
# hyperpriors
for (i in 1:ndiet) {
    alpha.mean[i] ~ dnorm(mu.alpha, prec.alpha) #
        parameter of interest
    beta.mean[i] ~ dnorm(mu.beta, prec.beta)    #
        parameter of interest
#   beta.mean[i] ~ dnorm(mu.beta, prec.beta)T(0,) #
    parameter of interest
```

```
      alpha.sig[i] ~ dunif(1.0E-6, 500)
      alpha.tau[i] ← 1/(alpha.sig[i] * alpha.sig[i])
      beta.sig[i] ~ dunif(1.0E-6, 500)
      beta.tau[i] ← 1/(beta.sig[i] * beta.sig[i])
}
# # # # # # # # # # # # # # # # # # # # # # # # # # # # # # # # # #
# hyper-hyperpriors

mu.alpha ~ dnorm(0.0,1.0E-6)
mu.beta ~ dnorm(0.0,1.0E-6)
#mu.beta ~ dnorm(0.0,1.0E-6)T(0,)
sig.alpha ~ dunif(1.0E-6, 1000)
prec.alpha ← 1/(sig.alpha * sig.alpha)
sig.beta ~ dunif(1.0E-6, 1000)
prec.beta ← 1/(sig.beta * sig.beta)

diff_a[1] ← alpha.mean[1]-alpha.mean[2]
diff_a[2] ← alpha.mean[1]-alpha.mean[3]
diff_a[3] ← alpha.mean[2]-alpha.mean[3]

diff_b[1]← beta.mean[1]-beta.mean[2]
diff_b[2]← beta.mean[1]-beta.mean[3]
diff_b[3] ← beta.mean[2]-beta.mean[3]
}

INITS
list(alpha1=c(0,0,0,0,0,0,0,0),
    beta1=c(0,0,0,0,0,0,0,0), alpha2=c(0,0,0,0),
    beta2=c(0,0,0,0), alpha3=c(0,0,0,0),
        beta3=c(0,0,0,0),
    alpha.mean = c(0,0,0), beta.mean  = c(0,0,0), tau
        = c(1,1,1),
    alpha.sig = c(1,1,1), beta.sig=c(1,1,1), mu.alpha
        = 0,
    mu.beta = 0, sig.alpha = 1, sig.beta = 1)


DATA
list(N1=88, N2=44, N3=44, ndiet = 3)

  y[]   x[] diet[] rat[]
  240   1    1 1
  250   8    1 1
  255  15    1 1
   .    .     .
   .    .     .
   .    .     .
  550  50    3 4
  553  57    3 4
  569  64    3 4
END
```

Once again, the model in Box 6.11 converged quickly, but since computing was fast and cheap we ran it for 50,000 iterations and discarded the first 25,000.

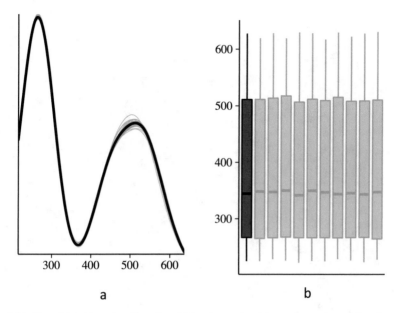

Fig. 6.16 Kernel densities (**a**) and boxplots (**b**) for observed weights and posterior predicted weights using data from full rat data

In order to check whether the observed data looked like a sample generated by the model, thereby indicating that the model was adequate, we used the code in Box 6.12 to generate posterior predictive samples from the model. Figure 6.16 displays kernel density estimates of the observed data (dark gray) and 25 randomly selected posterior predictive samples (light gray), and boxplots of the observed data (dark gray) and 10 randomly selected posterior predictive samples (light gray). The kernel densities of the posterior predictive samples are barely distinguishable from the observed data, and the boxplots also indicate very close agreement between the observed data and the posterior predictive samples. Hence we conclude that the model is adequate.

Code box 6.12 R code for generating posterior predictive distributions the full rat data set.

```
rm(list=ls())
#
#   Be sure to set working directory!!
#
setwd("   ")

library(nlme)
library("bayesplot")

head(BodyWeight)
D1 <- BodyWeight[BodyWeight[,4]==1,]
wt1 <- as.matrix(D1$weight)
t1 <- as.matrix(D1$Time)
r1 <- as.matrix(as.numeric(as.character(D1$Rat)))
```

```
n1 <- length(wt1)
diet1 <- rep(1,n1)

D2 <- BodyWeight[BodyWeight[,4]==2,]
wt2 <- as.matrix(D2$weight)
t2 <- as.matrix(D2$Time)
r2 <- as.matrix(as.numeric(as.character(D2$Rat)) - 8)
n2 <- length(wt2)
diet2 <- rep(2,n2)

D3 <- BodyWeight[BodyWeight[,4]==3,]
wt3 <- as.matrix(D3$weight)
t3 <- as.matrix(D3$Time)
r3 <- as.matrix(as.numeric(as.character(D3$Rat)) - 12)
n3 <- length(wt3)
diet3 <- rep(3,n3)

wt <- rbind(wt1,wt2,wt3)
wt <- as.vector(wt)
t <- rbind(t1,t2,t3)
r <- rbind(r1,r2,r3)
diet <- c(diet1,diet2,diet3)

rm(D1,D2,D3,diet1,diet2,diet3)

Nobs <- length(wt)
Nrats1 <- 8
Nrats2 <- 4
Nrats3 <- 4
Ndiets <- 3
Nperrat <- 11

k <- 25000   # k = posterior sample size

# Initialize matrices to hold BUGS output


alpha.mean <- matrix(0,nrow=k,ncol=Ndiets)
alpha.sig <- matrix(0,nrow=k,ncol=Ndiets)
alpha1 <- matrix(0,nrow=k,ncol=Nrats1)
alpha2 <- matrix(0,nrow=k,ncol=Nrats2)
alpha3 <- matrix(0,nrow=k,ncol=Nrats3)
beta.mean <- matrix(0,nrow=k,ncol=Ndiets)
beta.sig <- matrix(0,nrow=k,ncol=Ndiets)
beta1 <- matrix(0,nrow=k,ncol=Nrats1)
beta2 <- matrix(0,nrow=k,ncol=Nrats2)
beta3 <- matrix(0,nrow=k,ncol=Nrats3)

mu.alpha <- rep(0,k)
mu.beta <- rep(0,k)
sig.alpha <- rep(0,k)
sig.beta <- rep(0,k)
sigma <- matrix(0,nrow=k,ncol=Ndiets)
```

```
# read BUGS output

y <- read.table("full rat growth
   model.out",header=FALSE,row.names = NULL)
alpha.mean[,1] <- y[1:k,2]
alpha.mean[,2] <- y[(k+1):(2*k),2]
alpha.mean[,3] <- y[((2*k)+1):(3*k),2]
alpha.sig[,1] <- y[((3*k)+1):(4*k),2]
alpha.sig[,2] <- y[((4*k)+1):(5*k),2]
alpha.sig[,3] <- y[((5*k)+1):(6*k),2]
alpha1[,1] <- y[((6*k)+1):(7*k),2]
alpha1[,2] <- y[((7*k)+1):(8*k),2]
alpha1[,3] <- y[((8*k)+1):(9*k),2]
alpha1[,4] <- y[((9*k)+1):(10*k),2]
alpha1[,5] <- y[((10*k)+1):(11*k),2]
alpha1[,6] <- y[((11*k)+1):(12*k),2]
alpha1[,7] <- y[((12*k)+1):(13*k),2]
alpha1[,8] <- y[((13*k)+1):(14*k),2]
alpha2[,1] <- y[((14*k)+1):(15*k),2]
alpha2[,2] <- y[((15*k)+1):(16*k),2]
alpha2[,3] <- y[((16*k)+1):(17*k),2]
alpha2[,4] <- y[((17*k)+1):(18*k),2]
alpha3[,1] <- y[((18*k)+1):(19*k),2]
alpha3[,2] <- y[((19*k)+1):(20*k),2]
alpha3[,3] <- y[((20*k)+1):(21*k),2]
alpha3[,4] <- y[((21*k)+1):(22*k),2]

alpha <- array(0,dim=c(25000,8,3))


for (j in 1:k){
  alpha[j,1,1] <- alpha1[j,1]
  alpha[j,2,1] <- alpha1[j,2]
  alpha[j,3,1] <- alpha1[j,3]
  alpha[j,4,1] <- alpha1[j,4]
  alpha[j,5,1] <- alpha1[j,5]
  alpha[j,6,1] <- alpha1[j,6]
  alpha[j,7,1] <- alpha1[j,7]
  alpha[j,8,1] <- alpha1[j,8]

  alpha[j,1,2] <- alpha2[j,1]
  alpha[j,2,2] <- alpha2[j,2]
  alpha[j,3,2] <- alpha2[j,3]
  alpha[j,4,2] <- alpha2[j,4]

  alpha[j,1,3] <- alpha3[j,1]
  alpha[j,2,3] <- alpha3[j,2]
  alpha[j,3,3] <- alpha3[j,3]
  alpha[j,4,3] <- alpha3[j,4]
}

beta.mean[,1] <- y[((22*k)+1):(23*k),2]
beta.mean[,2] <- y[((23*k)+1):(24*k),2]
beta.mean[,3] <- y[((24*k)+1):(25*k),2]
```

```
beta.sig[,1] <- y[((25*k)+1):(26*k),2]
beta.sig[,2] <- y[((26*k)+1):(27*k),2]
beta.sig[,3] <- y[((27*k)+1):(28*k),2]
beta1[,1] <- y[((28*k)+1):(29*k),2]
beta1[,2] <- y[((29*k)+1):(30*k),2]
beta1[,3] <- y[((30*k)+1):(31*k),2]
beta1[,4] <- y[((31*k)+1):(32*k),2]
beta1[,5] <- y[((32*k)+1):(33*k),2]
beta1[,6] <- y[((33*k)+1):(34*k),2]
beta1[,7] <- y[((34*k)+1):(35*k),2]
beta1[,8] <- y[((35*k)+1):(36*k),2]
beta2[,1] <- y[((36*k)+1):(37*k),2]
beta2[,2] <- y[((37*k)+1):(38*k),2]
beta2[,3] <- y[((38*k)+1):(39*k),2]
beta2[,4] <- y[((39*k)+1):(40*k),2]
beta3[,1] <- y[((40*k)+1):(41*k),2]
beta3[,2] <- y[((41*k)+1):(42*k),2]
beta3[,3] <- y[((42*k)+1):(43*k),2]
beta3[,4] <- y[((43*k)+1):(44*k),2]

beta <- array(0,dim=c(25000,8,4))


for (j in 1:k){
  beta[j,1,1] <- beta1[j,1]
  beta[j,2,1] <- beta1[j,2]
  beta[j,3,1] <- beta1[j,3]
  beta[j,4,1] <- beta1[j,4]
  beta[j,5,1] <- beta1[j,5]
  beta[j,6,1] <- beta1[j,6]
  beta[j,7,1] <- beta1[j,7]
  beta[j,8,1] <- beta1[j,8]

  beta[j,1,2] <- beta2[j,1]
  beta[j,2,2] <- beta2[j,2]
  beta[j,3,2] <- beta2[j,3]
  beta[j,4,2] <- beta2[j,4]

  beta[j,1,3] <- beta3[j,1]
  beta[j,2,3] <- beta3[j,2]
  beta[j,3,3] <- beta3[j,3]
  beta[j,4,3] <- beta3[j,4]
}

mu.alpha[] <- y[((50*k)+1):(51*k),2]
mu.beta[] <- y[((51*k)+1):(52*k),2]

sig.alpha[] <- y[((52*k)+1):(53*k),2]
sig.beta[] <- y[((53*k)+1):(54*k),2]

sigma[,1] <- y[((54*k)+1):(55*k),2]
sigma[,2] <- y[((55*k)+1):(56*k),2]
sigma[,3] <- y[((56*k)+1):(57*k),2]
rm(y)
```

```
# read posterior samples

wrep1_1 <- matrix(0,nrow=k,ncol=Nperrat)
wrep1_2 <- matrix(0,nrow=k,ncol=Nperrat)
wrep1_3 <- matrix(0,nrow=k,ncol=Nperrat)
wrep1_4 <- matrix(0,nrow=k,ncol=Nperrat)
wrep1_5 <- matrix(0,nrow=k,ncol=Nperrat)
wrep1_6 <- matrix(0,nrow=k,ncol=Nperrat)
wrep1_7 <- matrix(0,nrow=k,ncol=Nperrat)
wrep1_8 <- matrix(0,nrow=k,ncol=Nperrat)

wrep2_1 <- matrix(0,nrow=k,ncol=Nperrat)
wrep2_2 <- matrix(0,nrow=k,ncol=Nperrat)
wrep2_3 <- matrix(0,nrow=k,ncol=Nperrat)
wrep2_4 <- matrix(0,nrow=k,ncol=Nperrat)

wrep3_1 <- matrix(0,nrow=k,ncol=Nperrat)
wrep3_2 <- matrix(0,nrow=k,ncol=Nperrat)
wrep3_3 <- matrix(0,nrow=k,ncol=Nperrat)
wrep3_4 <- matrix(0,nrow=k,ncol=Nperrat)

for (i in 1:k){
  mu1_1 <- alpha1[i,1] + beta1[i,1]*t[1:11]
  wrep1_1[i,] <-
      rnorm(n=Nperrat,mean=mu1_1,sd=sigma[i,1])
  mu1_2 <- alpha1[i,2] + beta1[i,2]*t[12:22]
  wrep1_2[i,] <-
      rnorm(n=Nperrat,mean=mu1_2,sd=sigma[i,1])
  mu1_3 <- alpha1[i,3] + beta1[i,3]*t[23:33]
  wrep1_3[i,] <-
      rnorm(n=Nperrat,mean=mu1_3,sd=sigma[i,1])
  mu1_4 <- alpha1[i,4] + beta1[i,4]*t[34:44]
  wrep1_4[i,] <-
      rnorm(n=Nperrat,mean=mu1_4,sd=sigma[i,1])
  mu1_5 <- alpha1[i,5] + beta1[i,5]*t[1:11]
  wrep1_5[i,] <-
      rnorm(n=Nperrat,mean=mu1_5,sd=sigma[i,1])
  mu1_6 <- alpha1[i,6] + beta1[i,6]*t[12:22]
  wrep1_6[i,] <-
      rnorm(n=Nperrat,mean=mu1_6,sd=sigma[i,1])
  mu1_7 <- alpha1[i,7] + beta1[i,7]*t[23:33]
  wrep1_7[i,] <-
      rnorm(n=Nperrat,mean=mu1_7,sd=sigma[i,1])
  mu1_8 <- alpha1[i,8] + beta1[i,8]*t[34:44]
  wrep1_8[i,] <-
      rnorm(n=Nperrat,mean=mu1_8,sd=sigma[i,1])

  mu2_1 <- alpha2[i,1] + beta2[i,1]*t[1:11]
  wrep2_1[i,] <-
      rnorm(n=Nperrat,mean=mu2_1,sd=sigma[i,2])
  mu2_2 <- alpha2[i,2] + beta2[i,2]*t[12:22]
```

```
wrep2_2[i,] <-
    rnorm(n=Nperrat,mean=mu2_2,sd=sigma[i,2])
mu2_3 <- alpha2[i,3] + beta2[i,3]*t[23:33]
wrep2_3[i,] <-
    rnorm(n=Nperrat,mean=mu2_3,sd=sigma[i,2])
mu2_4 <- alpha2[i,4] + beta2[i,4]*t[34:44]
wrep2_4[i,] <-
    rnorm(n=Nperrat,mean=mu2_4,sd=sigma[i,2])

mu3_1 <- alpha3[i,1] + beta3[i,1]*t[1:11]
wrep3_1[i,] <-
    rnorm(n=Nperrat,mean=mu3_1,sd=sigma[i,3])
mu3_2 <- alpha3[i,2] + beta3[i,2]*t[12:22]
wrep3_2[i,] <-
    rnorm(n=Nperrat,mean=mu3_2,sd=sigma[i,3])
mu3_3 <- alpha3[i,3] + beta3[i,3]*t[23:33]
wrep3_3[i,] <-
    rnorm(n=Nperrat,mean=mu3_3,sd=sigma[i,3])
mu3_4 <- alpha3[i,4] + beta3[i,4]*t[34:44]
wrep3_4[i,] <-
    rnorm(n=Nperrat,mean=mu3_4,sd=sigma[i,3])

}


wrep <- cbind(wrep1_1,wrep1_2,wrep1_3,wrep1_4,
              wrep1_5,wrep1_6,wrep1_7,wrep1_8,
              wrep2_1,wrep2_2,wrep2_3,wrep2_4,
              wrep3_1,wrep3_2,wrep3_3,wrep3_4)

# head(wrep)

color_scheme_set("brightblue")
ind <- sample(1:k, 25, replace=FALSE)
ppc_dens_overlay(wt, wrep[ind, ])
ind2 <- sample(1:k, 10, replace=FALSE)
ppc_boxplot(wt, wrep[ind2, ],notch=FALSE)
```

As mentioned earlier, in a study such as this the parameters of ultimate interest are usually the group-specific parameters, i.e., $\mu_{\alpha i}$ and $\mu_{\beta i}$. These parameters are the mean intercepts ($\mu_{\alpha i}$) and mean slopes ($\mu_{\beta i}$) for the three diets. Any differences among the diets should manifest themselves as differences among these parameters. The marginal posterior densities for $\mu_{\alpha i}$ and $\mu_{\beta i}$, $i = 1, 2, 3$, are shown in Fig. 6.17. The most striking feature of the marginal posterior densities is probably the difference in scale between the densities of parameters for diet 1 and those for diets 2 and 3. This is due to sample size. We had 8 rats for diet 1 and 4 apiece for diets 2 and 3. Hence we learned much more about diet 1 than we did about the others.

The mean, standard deviation, 0.025 quantile, and 0.975 quantile from the marginal posterior samples for the parameters are in Table 6.3. Both Fig. 6.17 and Table 6.3 indicate that $\mu_{\alpha 1}$ may be different from $\mu_{\alpha 2}$ and $\mu_{\alpha 3}$, but they also suggest that there is probably not a significant difference among the $\mu_{\beta i}$'s.

This is further confirmed by examining the marginal posterior samples of ($\mu_{\alpha 1} - \mu_{\alpha 2}$), ($\mu_{\alpha 1} - \mu_{\alpha 3}$), ($\mu_{\alpha 2} - \mu_{\alpha 3}$), ($\mu_{\beta 1} - \mu_{\beta 2}$), ($\mu_{\beta 1} - \mu_{\beta 3}$), and ($\mu_{\beta 2} - \mu_{\beta 3}$). These

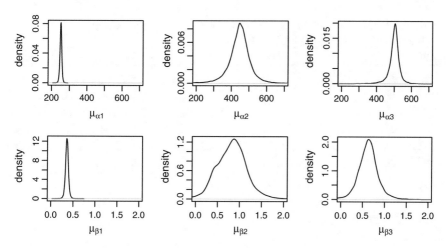

Fig. 6.17 Marginal posterior distributions for $\mu_{\alpha i}$ and $\mu_{\beta i}$, $1 = 1, 2, 3$

Table 6.3 Mean, standard deviation, 0.025 quantiles, and 0.975 quantiles from marginal posterior densities for $\mu_{\alpha i}$ and $\mu_{\beta i}$, $i = 1, 2, 3$, using the full rat data set

Parameter	Mean	Std dev	0.025	0.975
$\mu_{\alpha 1}$	251.6	5.7	240.2	263.3
$\mu_{\alpha 2}$	445.8	66.0	305.3	574.5
$\mu_{\alpha 3}$	498.9	36.9	415.7	562.1
$\mu_{\beta 1}$	0.36	0.04	0.29	0.44
$\mu_{\beta 2}$	0.82	0.39	0.11	1.55
$\mu_{\beta 3}$	0.62	0.28	0.11	1.15

are readily available by computing the appropriate quantity from the joint posterior sample produced by **OpenBUGS**. The 0.025 and 0.975 quantiles for the differences are shown in Table 6.4. These constitute approximate 95% credible intervals. The credible intervals for $(\mu_{\alpha 1} - \mu_{\alpha 2})$ and $(\mu_{\alpha 1} - \mu_{\alpha 3})$ do not include 0, indicating that 0 is not a reasonable value for the mean of the these differences, and hence $\mu_{\alpha 1}$ is significantly different from $\mu_{\alpha 2}$ and $\mu_{\alpha 3}$ at the 0.05 level of significance. All of the other credible intervals include 0, and hence there are no significant differences among the $\mu_{\beta i}$'s, $i = 1, 2, 3$, and $\mu_{\alpha 2}$ is not significantly different from $\mu_{\alpha 3}$.

Table 6.4 0.025 quantiles, and 0.975 quantiles from marginal posterior densities of differences among $\mu_{\alpha i}$ and $\mu_{\beta i}$, $i = 1, 2, 3$ for full rat data set

Difference	0.025	0.975
$\mu_{\alpha 1} - \mu_{\alpha 2}$	−323.9	−53.2
$\mu_{\alpha 1} - \mu_{\alpha 3}$	−311.3	−164.5
$\mu_{\alpha 2} - \mu_{\alpha 3}$	−207.7	98.5
$\mu_{\beta 1} - \mu_{\beta 2}$	−1.20	0.25
$\mu_{\beta 1} - \mu_{\beta 3}$	−0.80	0.25
$\mu_{\beta 2} - \mu_{\beta 3}$	−0.65	1.09

6.3 Exercises

The file snakes_data.txt available online contains observations on mass (Y) and length (X) of snakes for each of 4 populations of interest (1, 2, 3, and 4).

1. Use **OpenBUGS** to fit the following linear model to the data in each group (i.e., analyze the data from each group separately):

$$y_i \mid \mu_i, \sigma^2 \sim \mathbf{N}(\mu_i, \sigma^2) \quad \mu_i = \alpha + \beta x_i$$
$$\alpha \sim \mathbf{N}(0, 10000); \quad \beta \sim \mathbf{N}(0, 10000) \quad \sigma^2 \sim \mathbf{Ga}^{-1}(0.001, 0001)$$
$$i = 1, 2, \ldots, n$$

2. Another way to model each population separately is to use *dummy* variables and fit one model to *all* the data simultaneously.

    Let $z_{1i} = 1$ if observation i is from group 1 and 0 if it isn't. Similarly, let $z_{2i} = 1$ if observation i is from group 2 and 0 if not, $z_{3i} = 1$ if observation i is from group 3 and 0 if not, and $z_{4i} = 1$ if observation i is from group 4 and 0 if not.

    Use **OpenBUGS** the following model:

$$y_i \mid \mu_i, \sigma^2 \sim \mathbf{N}(\mu_i, \sigma_i^2)$$
$$\mu_i = z_{1i}(\alpha_1 + \beta_1 x_i) + z_{2i}(\alpha_2 + \beta_2 x_i) + z_{3i}(\alpha_3 + \beta_3 x_i) + z_{4i}(\alpha_4 + \beta_4 x_i)$$
$$\sigma_i^2 = z_{1i} * \sigma_1^2 + z_{2i} * \sigma_2^2 + z_{3i} * \sigma_3^2 + z_{4i} * \sigma_4^2$$
$$\alpha_j \sim \mathbf{N}(0, 10000) \quad \beta_j \sim \mathbf{N}(0, 10000) \quad \sigma_j^2 \sim \mathbf{Ga}^{-1}(0.001, 0001)$$
$$i = 1, 2, \ldots, n_j \quad j = 1, 2, 3, 4$$

    Verify that the variances, intercepts, and slopes for each population are the same as those obtained in exercise 1.

3. Derive a 95% prediction interval for the mass of a snake with length = 50 from population 1.

4. Use **OpenBUGS** to fit the following hierarchical model to the snake mass data from exercise 1:

$$y_{ij} \mid \mu_{ij}, \sigma_i^2 \sim \mathbf{N}(\mu_{ij}, \sigma_i^2) \quad \mu_{ij} = \alpha_i + \beta_i x_j$$
$$\alpha_i \sim \mathbf{N}(\mu_\alpha, \delta_\alpha^2) \quad \beta_i \sim \mathbf{N}(\mu_\beta, \delta_\beta^2) \quad \sigma_i^2 \sim \mathbf{Ga}^{-1}(0.001, 0001)$$
$$\mu_\alpha \sim \mathbf{N}(0, 10000) \quad \mu_\beta \sim \mathbf{N}(0, 10000)$$
$$\delta_\alpha \sim \mathbf{Unif}(0, 100) \quad \delta_\beta \sim \mathbf{Unif}(0, 100)$$
$$j = 1, 2, \ldots, n_i \quad i = 1, 2, 3, 4$$

5. Compare *DIC*, *WAIC* and *LOO* for the models in exercises 2 and 4. Which model do you prefer?

References

Avery, T. E., Burkhart, H. E., & Bullock, B. P. (2019). *Forest Measurements* (6th ed.). Long Grove, IL: Waveland Press.

Draper, N. R., & Smith, H. (1998). *Applied Regression Analysis* (3rd ed.). New York, NY: Wiley.

Edwards, J. W., & Jannink, J. L. (2006). Bayesian modelling of heterogeneous error and genotype x environment interaction variances. *Crop Science, 46*, 820–833.

Gelfand, A. E., Hills, S. E., Racine-Poon, A., & Smith, A. F. M. (1990). Illustration of Bayesian inference in normal data models using Gibbs sampling. *Journal of the American Statistical Association, 85*(412), 972–985.

Gelman, A. (2006). Prior distributions for variance parameters in hierarchical models. *Bayesian Analysis, 1*(3), 515–533.

Green, E. J., & Valentine, H. T. (1998). Bayesian analysis of the linear model with heterogeneous variance. *Forest Science, 44*(1), 134–138.

Chapter 7
General Linear Models

Oftentimes, researchers are confronted with a situation in which an observed dependent variable, which may or may not be related to observed covariates, is not amenable to analysis via "usual" linear model methodology. For instance, the dependent variable may be a count of some phenomenon, e.g., the number of individuals per plot. Since counts are discrete, they clearly fail the usual normality assumption for dependent variables, which among other things, specifies the dependent variable is continuous. Or perhaps the dependent variable is a proportion, constrained to lie in the interval (0, 1), e.g., the proportion of habitable land in a given area. The latter also fails the normality assumption, which specifies that the dependent variable is defined on the interval $(-\infty, \infty)$.

The solution in situations such as these may be to use General Linear Models (GLMs). The basic idea underlying GLMs is to use a function, say $h(\cdot)$, to transform a linear function of covariates (say $\mathbf{X}\beta$), where $\mathbf{X}$ is a matrix of covariates and β is a vector of regression coefficients, in such a way that the transformed variable $h(\mathbf{X}\beta)$ satisfies the usual regression assumptions. This function is often called the *link function*. Predicted values of Y can be obtained via the inverse of the link function, $h^{-1}(\cdot)$.

In this chapter we will consider two common GLMs: Poisson regression and logistic regression. They differ in the characteristics of the observed variables (Y) and in the link functions used.

7.1 Poisson Regression

The Poisson model is the default model for count data. Recall the Poisson distribution:

$$f(y \mid \lambda) = \frac{\lambda^y e^{-\lambda}}{y!}, \quad y = 0, 1, 2, \ldots; \lambda > 0. \qquad (7.1.1)$$

© Springer Nature Switzerland AG 2020
E. J. Green et al., *Introduction to Bayesian Methods in Ecology and Natural Resources*,
https://doi.org/10.1007/978-3-030-60750-0_7

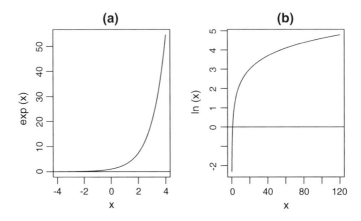

Fig. 7.1 Plot of **a** exp(x) versus x, and **b** ln(x) versus x

As mentioned in Chap. 2, the parameter λ is both the mean and the variance of the Poisson distribution. An important property of the Poisson distribution is the restriction of λ to the interval $(0, \infty)$, i.e., λ must be positive. This has important ramifications if we try to model the mean of the dependent variable as a function of one or more covariates. Suppose we observe n independent realizations of $(Y, X_1, X_2, \ldots, X_p)$, i.e., we have p covariates. We *must* ensure that the model always yields positive values. One way to guarantee this is to use the link function $ln(\cdot)$,[1] and let

$$y_j \propto Pois(\theta_j), \tag{7.1.2}$$
$$\theta_j = exp(\beta_0 + \beta_1 x_{1j} + \beta_2 x_{2j} + \cdots + \beta_p x_{pj}), \tag{7.1.3}$$

where y_j is the jth value of the dependent variable Y; x_{ij} is the value of covariate X_i observed at observation j; β_i is the unknown value of the coefficient associated with X_i; $i = 1, 2, \ldots, p$, $j = 1, 2, \ldots, n$. Using the link function $ln(\cdot)$ ensures that θ_j, the mean of y_j, will be positive (see Fig. 7.1).

7.1.1 Poisson Regression Example

This example displays how a basic Poisson regression model can be used to assess the effect of covariates on avian abundance. The data come from a large breeding landbird monitoring program implemented across nine national parks in the northeastern United States by the Northeast Temperate Inventory and Monitoring Network (NETN) of the National Park Service to assess trends in avian abundance across a

---

[1]The link function is $ln(\cdot)$ because we think of the link function h so that $h(\mu) = X\beta$ and $\mu = h^{-1}(X\beta)$, where $h^{-1}(\cdot)$ is the inverse function. Hence in Poisson regression, $ln(\cdot)$ is the link function and $exp(\cdot)$ is the inverse link function.

thirteen year period and understand how different forest and landscape characteristics were driving these changes. In this example we focus on bird count data collected at 51 sites in Acadia National Park in Maine. At each of the 51 sites, volunteer observers performed a 10-min. point count survey in which they recorded each bird they saw or heard within a 250 m radius. Here we seek to understand the effect of three covariates on the total number of individual birds detected at each point: the percent forest occurring within a 1 km radius of the point count location, the amount of forest regeneration (used to assess the effect of deer browsing, measured as the number of saplings and seedlings >0.5 m tall per ha), and the amount of tree basal area (m^2/ha).

Let y_j = the bird count at point j, x_{1j} = the basal area at point j, x_{2j} = the percent forest at point j, and x_{3j} = the amount of regeneration at point j. We posit the following logistic regression model with vague priors:

$$y_j \propto \textbf{Pois}(\lambda_j), \tag{7.1.4}$$

$$\lambda_j = exp(\beta_0 + \beta_1 x_{1j} + \beta_2 x_{2j} + \beta_3 x_{3j}), \ j = 1, 2, \ldots, 51, \tag{7.1.5}$$

$$\beta_i \sim \textbf{N}(0, 1.0 \times 10^6), \ i = 0, 1, 2, 3. \tag{7.1.6}$$

The model represented by (7.1.4)–(7.1.6) is surprisingly simple to code in **Open-BUGS**, as shown in Box 7.1. The full data set is not shown in the Box, but is available online. Also, two steps are required to load the data. First, highlight the command list(N=51) and load it, and then highlight the data in rectangular format and load it. The covariate data in Box 7.1 are "centered," i.e., we present $x_{ij} - \bar{x}_i$ rather than $x_{ij}, i = 1, 2, 3, j = 1, 2, \ldots, 51.$[2] This is done because, as mentioned in Lunn et al. (2013), high posterior correlation among regression parameters may cause convergence issues in **OpenBUGS**. Centering the data reduces this correlation.

Code box 7.1 OpenBUGS code for fitting Poisson regression to avian data.

```
MODEL
{
  # Poisson regression

  # y=bird count, x1=basal area, x2 = pct forest,
  # x3=regeneration
  for (j in 1:N) {
    log(lambda[j]) ← beta.0 + beta.1 * x1[j] +
    beta.2 * x2[j] + beta.3 * x3[j]
    y[j] ~ dpois(lambda[j])
  }
  # Priors -------------------------------------------
  beta.0 ~ dnorm(0, 1.0E-6)
  beta.1 ~ dnorm(0, 1.0E-6)
  beta.2 ~ dnorm(0, 1.0E-6)
  beta.3 ~ dnorm(0, 1.0E-6)
}
INITS
```

---

[2]This is also sometimes referred to as "correcting" the x_i values.

```
list(beta.0=1,  beta.1=0,  beta.2=0,  beta.3=0 )
DATA
list(N =51)

 y[]      x1[]       x2[]       x3[]
15   0.0297845690674945   -0.756601405272028
     0.176921476886089
13   0.0141927691910787   -0.117398629041489
     0.376319751615821
13   -0.635286001298159   0.438488570743193
     -0.150047675842499
        .        .        .
        .        .        .
        .        .        .
13   -0.383339062053776   1.25634180283572
     0.331506120902055
 7   0.57341843544209    1.27803244412902
     -0.151536075286571
 6   -0.754800681810292   1.15806754489488
     -0.419134627968263
END
```

We ran the model in Box 7.1 for 50,000 iterations and discarded the first 25,000 (quantile and trace plots showed that the sampler converged in under 2500 iterations). The 0.025, 0.500, and 0.975 quantiles from iterations 25,001 through 50,000 are presented for each of the model parameters in Table 7.1. The 0.025 and 0.975 quantiles define an approximate 95% credible interval. The credible intervals for β_1 and β_3 both overlap 0, so we conclude that there is no significant effect of basal area or regeneration on observed avaian abundance. To check this, we compared the DIC of the model in Box 7.1 to the DIC of the model *without* basal area or regeneration included. The former was 242.0, while the latter was 239.0, indicating that the model without regeneration and basal area was supported by the data more than the model with these covariates. The marginal posterior distributions of the parameters in the reduced model ($\lambda_j = exp(\beta_0 + \beta_1 x_{1j})$) are shown in Fig. 7.2. The posterior mean for β_1 was -0.138, indicating a negative effect of percent forest on observed avian abundance.

Table 7.1 0.025, 0.500 (median) and 0.975 quantiles for parameters in Poisson regression fitted to avian data

Parameter	0.025	0.5	0.975
β_0	2.430	2.556	2.676
β_1	-0.220	-0.067	0.083
β_2	-0.280	-0.144	-0.005
β_3	-0.219	0.060	0.323

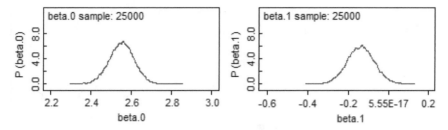

Fig. 7.2 Marginal posterior distributions from Poisson regression fit to avian data

It is important to note that here we are directly modeling covariate effects on the *observed* avian abundance. It is highly likely that only a subset of the birds at the study sites were counted (i.e., detection was not perfect) and we would want to account for this imperfect detection in a more sophisticated modeling exercise.

7.2 Logistic Regression

Logistic regression is useful when the response variable is constrained to the interval $(0, 1)$. For instance, suppose we are interested in presence/absence of a species in a given area, and suppose the response variable Y is coded as 1 or 0, denoting presence (1) or absence (0). Further suppose that a covariate X is thought to be related to Y. We might naively regress Y on X using ordinary regression procedures, but this would soon lead to difficulties as predicted variables would not be constrained to take on the value 0 or 1, and, perhaps even worse, could fall outside the interval $(0, 1)$. This may be avoided by modeling the response as a Bernoulli random variable. As mentioned in Sect. 2.5.1, a Bernoulli random variable is similar to a Binomial random variable, except that for a Bernoulli we observe only one trial and the two possible outcomes of the trial are 1 with probability p, or 0 with probability $(1 - p)$.

The parameter of interest is therefore the parameter p. Although p is defined on the unit interval, the *logit* of p, as defined by the logistic function, is defined on the interval $(-\infty, \infty)$. Let's call u the logit of p. Then,

$$u = logit(p) = log\left(\frac{p}{1-p}\right). \tag{7.2.1}$$

Note that the logit of p is simply the log of the odds of p; this is often referred to as the log-odds of p. Some simple algebra shows that the *inverse logit* is

$$p = logit^{-1}(u) = \frac{exp(u)}{1 + exp(u)}, \tag{7.2.2}$$

Fig. 7.3 Plot of logit(p)
versus p

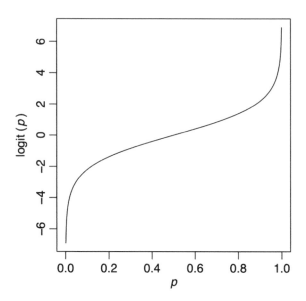

where, as usual, $exp(u) = e^u$. Figure 7.3 displays a plot of logit(p) versus p. As shown in the figure, logit(p) is defined over $(-\infty, \infty)$ while p is defined over $(0, 1)$.

Given the respective ranges of p and logit(p), it makes sense to explore the relationship of logit(p) with potential covariates. Once we have a predicted logit(p), we can recover p via the inverse logit transformation (Eq. 7.2.2).

While the logit is arguably the most common link function for dealing with data constrained to the interval $(0,1)$, there are other functions that could also be considered, e.g., the *probit* or the *complimentary log-log* (cloglog) link functions. We will concentrate on the *logistic* link in this chapter, but in practice we recommend exploring several link functions and choosing the one which yields the lowest DIC, WAIC, and or LOO, as illustrated in Carlin and Louis (2009).

7.2.1 Bernoulli Logistic Example

Suzuki et al. (2006) studied the endangered burrowing spider *Lycosa ishikariana* in Japan. Among other things, they recorded the presence or absence of the spider on 39 beaches. For 28 of the beaches they also presented the median sand grain size (*mm*).[3] We will restrict our attention to the 28 beaches for which he median grain size was reported. These data are presented in Table 7.2.

We might wish to investigate whether the probability of spider presence is related to median grain size. Let $y_i = 1$ if spiders are present on beach i and 0 if not; p_i

---

[3]This data is also used in McDonald (2014).

Table 7.2 Spider presence (1) or absence (0) and sand grain size (*mm*), for 28 beaches (Suzuki et al. 2006)

Presence/absence	Grain size	Presence/absence	Grain size
0	0.245	0	0.432
0	0.247	1	0.473
1	0.285	1	0.509
1	0.299	1	0.529
1	0.327	0	0.561
1	0.347	0	0.569
0	0.356	1	0.594
1	0.360	1	0.638
0	0.363	1	0.656
1	0.364	1	0.816
0	0.398	1	0.853
1	0.400	1	0.938
0	0.409	1	1.036
1	0.421	1	1.045

= probability of spider presence on beach i; x_i = median grain size on beach i; $i = 1, 2, \ldots, n$; and n = number of beaches studied. Then a plausible model is:

$$y_i \sim \textbf{Bern}(p_i), \tag{7.2.3}$$

$$logit(p_i) = \alpha + \beta x_i, \tag{7.2.4}$$

where $y_i \sim \textbf{Bern}(p_i)$ indicates that y_i follows a Bernoulli distribution with parameter p_i. Note that although logit(p_i) is a linear function of x_i, we do not include the usual normal error term (e_i) we are accustomed to seeing with linear models (e.g., see Eq. 6.0.1). This is because in the usual linear model, the error term represents observational error; it accounts for the fact that the observed values of the dependent variable differ from their means. However, in the present case we do not actually observe the p_i values, so it would be inappropriate to include an observational error term. Here, the "data model" is $y_i \sim \textbf{Bern}(p_i)$, and this model includes the mean and variance of y_i. They are p_i and $p_i(1 - p_i)$, respectively.[4]

All that remains to complete our Bayesian model is to specify prior distributions for α and β. As usual, we choose the vague priors $\alpha \sim \textbf{N}(0, 1 \times 10^6)$ and $\beta \sim \textbf{N}(0, 1 \times 10^6)$.

The **OpenBUGS** code to fit the logistic regression model to the spider presence/absence data is in Box 7.2. Note that in the code, we altered the *logit* model; instead of fitting $logit(p_i) = \alpha + \beta x_i$, we fitted $logit(p_i) = \alpha + \beta(x_i - \bar{x})$, i.e., we centered the covariate data (as we did in Sect. 7.1.1).

---

[4]The mean and variance for a Bernoulli random variable follow from the binomial distribution with $n = 1$.

Code box 7.2 OpenBUGS code for fitting logistic regression to spider presence/absence data.

```
# Spider presence/absence model with covariate
# covariate is grain size
Model
{
for (i in 1:n) {
    y[i]~ dbern(p[i])
    logit(p[i]) ← alpha + beta*(x[i] - mean(x[ ]))
    }
alpha ~ dnorm(0,1.0E-6)
beta ~ dnorm(0,1.0E-6)
}
Data
list(n=28,
x=c(0.245, 0.247, 0.285, 0.299, 0.327, 0.347, 0.356,
    0.360, 0.363, 0.364, 0.398, 0.400, 0.409, 0.421,
    0.432, 0.473, 0.509, 0.529, 0.561, 0.569, 0.594,
    0.638, 0.656, 0.816, 0.853, 0.938 ,1.06, 1.045),
y = c(0, 0, 1, 1, 1, 1, 0, 1, 0, 1, 0, 1, 0, 1, 0, 1,
    1, 1, 0, 0, 1, 1, 1, 1, 1, 1, 1, 1))
Inits
list(alpha=0, beta = 1)
```

Quantile and history plots of the nodes alpha and beta revealed that the model in Box 7.2 converged quickly. However computation was cheap, so we ran the model for 50,000 iterations and computed DIC, WAIC, and LOO from iterations 25,001 through 50,000. The R code to compute WAIC and LOO are in Box 7.3.

Code box 7.3 R code for computing WAIC and LOO for OpenBUGS model in Box 7.2.

```
rm(list=ls())
#  Be sure to set directory to path containing
#  OpenBUGS posterior sample
setwd(" ")
library("loo")
library("boot")
x <- c(0.245, 0.247, 0.285, 0.299, 0.327, 0.347, 0
    .356,
    0.360, 0.363, 0.364, 0.398, 0.400, 0.409, 0.421,
    0.432, 0.473, 0.509, 0.529, 0.561, 0.569, 0.594,
    0.638, 0.656, 0.816, 0.853, 0.938, 1.036, 1.045)
y <- c(0, 0, 1, 1, 1, 1, 0, 1, 0, 1, 0, 1, 0, 1, 0, 1,
    1, 1, 0, 0, 1, 1, 1, 1, 1, 1, 1, 1)
xbar <- mean(x)
corr_x <- x-xbar
Nobs <- length(y)
k <- 25000   # k = posterior sample size
# Initialize matrices to hold OpenBUGS output
alpha <- matrix(0,nrow=k,ncol=1)
beta <- matrix(0,nrow=k,ncol=1)
# read OpenBUGS output
z <- read.table("spider.out",header=FALSE,row.names =
    NULL)
alpha[,1] <- z[1:k,2]
```

Table 7.3 DIC, WAIC, and LOO for spider model with and without covariate (median grain size)

Model	DIC	WAIC	LOO
With covariate	34.7	35.0	35.0
Without covariate	37.2	37.3	37.3

```
beta[,1] <- z[(k+1):(2*k),2]
# initialize log-likelihood matrices
log_lik <- matrix(0,nrow=k,ncol=Nobs)
# Compute loglikelihood for each observation and each
    value in
# joint posterior sample
for (i in 1:k){
  # parameters have index i
  for (j in 1:Nobs){
    # observations have index j
    p <- inv.logit(alpha[i]+beta[i]*corr_x[j])
    log_lik[i,j] <-
        dbinom(y[j],size=1,prob=p,log=TRUE)
  }
}
# Get LOO and WAIC from loo
LOO <- loo(log_lik); LOO
WAIC <- waic(log_lik); WAIC
```

The resulting DIC, WAIC, and LOO values are presented in Table 7.3. It may be interesting to investigate whether the covariate median grain size improved the model. To this end we fitted the model

$$y_i \sim Bernoulli(p_i), \tag{7.2.5}$$

$$logit(p_i) = \alpha, \tag{7.2.6}$$

$$\alpha \sim \mathbf{N}(0, 1.0 \times 10^6). \tag{7.2.7}$$

The OpenBUGS code for model (7.2.5) is in Box 7.4 and the R code for computing DIC, WAIC and LOO is in Box 7.5. The resulting values are in Table 7.3. Although the metrics in Table 7.3 slightly favor the model with the covariate, the differences are not substantial and do not firmly establish which model is superior. For the rest of this section we choose to use the model with the covariate, but we could just as easily have chosen the model without it.

Code box 7.4 OpenBUGS code for fitting Bernoulli model to spider presence/absence data without covariate.

```
# Spider presence/absence model without covariate
Model
{
for (i in 1:n) {
    y[i]~ dbern(p[i])
    logit(p[i]) <- alpha
```

```
      }
alpha  ~  dnorm (0 ,1.0E -6)
}
Data
list (n=28 ,
y = c (0,  0,  1,  1,  1,  1,  0,  1,  0,  1,  0,  1,  0,  1,  0,  1,
      1,  1,  0,  0,  1,  1,  1,  1,  1,  1,  1,  1))
Inits
list (alpha =1)
```

Code box 7.5 R code for computing WAIC and LOO for OpenBUGS model in Box 7.4.
```
rm (list =ls ())
#   Be sure to set directory to path containing
#   OpenBUGS posterior sample
setwd (" ")
library ("loo")
library ("boot")
y ← c (0,  0,  1,  1,  1,  1,  0,  1,  0,  1,  0,  1,  0,  1,  0,  1,
    1,
      1,  0,  0,  1,  1,  1,  1,  1,  1,  1,  1)
Nobs ← length (y)
k ← 50000   # k = posterior sample size
# Initialize matrices to hold OpenBUGS output
alpha ← matrix (0 ,nrow=k ,ncol =1)
# read OpenBUGS output
z ← read.table ("spider.out ",header =FALSE ,row.names =
    NULL)
alpha [,1] ← z [1:k ,2]
# initialize log-likelihood matrices
log_lik ← matrix (0 ,nrow=k ,ncol =Nobs)
# Compute loglikelihood for each observation and each
    value in
# joint posterior sample
for (i in 1:k){
  # parameters have index i
  for (j in 1:Nobs){
    # observations have index j
    p ← inv.logit (alpha [i])
    log_lik [i,j] ← dbinom (y[j],size =1,prob=p,log =TRUE)
  }
}
# Get LOO and WAIC from loo
LOO ← loo (log_lik); LOO
WAIC ← waic (log_lik); WAIC
```

The marginal posterior densities for α and β are displayed in Fig. 7.4. A plot of the fitted logistic regression model and the raw data is presented in Fig. 7.4. The R code for generating Fig. 7.5 is in Box 7.6. Note that the fitted model in the figure was the obtained using the posterior means of α and β. Plots such as Fig. 7.4 are different from the usual plots of data and fitted models because in presence/absence models, the data are a series of 0's and 1's and hence the fitted model never seems to "go through" the observed data as it does for the usual linear model (e.g., see Fig. 6.7).

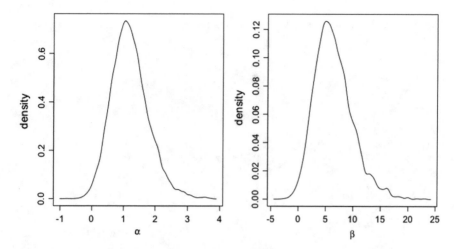

Fig. 7.4 Marginal posterior densities from fitting logistic regression to spider data

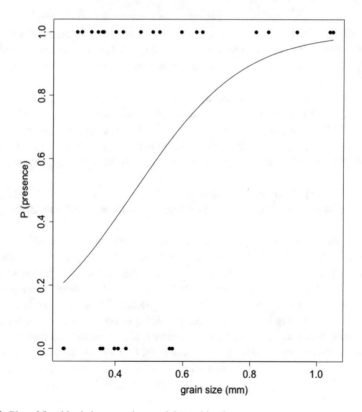

Fig. 7.5 Plot of fitted logistic regression model to spider data

Code box 7.6 R code for plotting fitted logistic regression against spider presence/absence data.

```
rm(list=ls())
par(mfrow=c(1,1))
x <- c(0.245, 0.247, 0.285, 0.299, 0.327, 0.347, 0
    .356,
    0.360, 0.363, 0.364, 0.398, 0.400, 0.409, 0.421,
    0.432, 0.473, 0.509, 0.529, 0.561, 0.569, 0.594,
    0.638, 0.656, 0.816, 0.853, 0.938, 1.036 ,1.045)
y <- c(0, 0, 1, 1, 1, 1, 0, 1, 0, 1, 0, 1, 0, 1, 0, 1,
    1, 1, 0, 0, 1, 1, 1, 1, 1, 1, 1, 1)
xgrid <- seq(from=min(x), to=max(x),length=101)
corr_x <- xgrid-mean(xgrid)
alpha <- 1.177
beta <- 6.294
p_logit <- exp(alpha + beta*corr_x)/(1+exp(alpha +
    beta*corr_x))
plot(xgrid,p_logit,xlab="grain size
    (mm)",ylab="P(presence)",
pch=20,type = "l",ylim=c(0,1))
points(x,y,pch=19)
```

OpenBUGS includes several convenient ways to visualize the results of fitting Bayesian logistic regressions. These are available by clicking on the `Compare` button under the `Inference` tool, selecting the parameter to investigate, and the iterations to use.[5]

Box plots for the posterior distribution of p_i are shown in Fig. 7.6 for each beach $(i = 1, \ldots, n)$. The baseline in the figure is the global mean of the $n = 28$ posterior means. From the figure, we see that there are several beaches for which the probability of spider presence is reasonably high (those on the right side of the figure) but there do not appear to be any where the probability of presence is low (or the probability of absence is high).

In Fig. 7.7 we present posterior density strips for each beach. Darker areas in the strips correspond to higher probability. Figure 7.7 tells basically the same story as Fig. 7.6.

Finally, we calculated the posterior probability of the model correctly classifying each of the 28 beaches. This was done by computing the posterior p_i for each beach $(i = 1, \ldots, n)$ for each pair of (α_j, β_j) values in the joint posterior sample $(j = 1, \ldots, 25000)$. If the predicted probability was ≥ 0.5, spiders were predicted to be present for that beach, if not they were predicted to be absent. If the beach did indeed have spiders and spiders were predicted to be present, this was a correct classification. Similarly, is the beach did not have spiders, and the beach was predicted to not have spiders, the classification was correct. Otherwise, the classification was incorrect. We then computed the posterior probability of a correct classification for each beach by averaging over the joint posterior sample. The R code for this is in Box 7.7.

---

[5]Of course, the parameter to investigate must have been selected to monitor prior to updating the model.

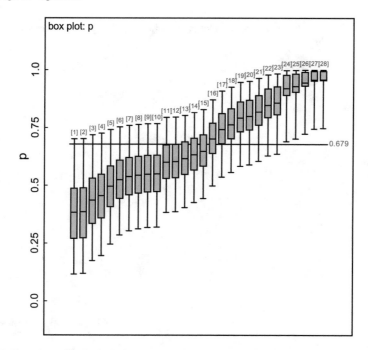

Fig. 7.6 Box plots of p_i for logistic regression fitted to spider data

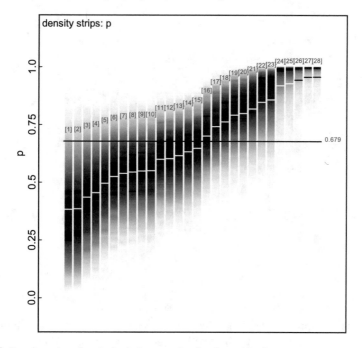

Fig. 7.7 Density strips of p_i for logistic regression fitted to spider data

Code box 7.7 R code for computing and plotting posterior probability of correct classification for each beach in spider data.

```
rm(list=ls())
#  Be sure to set directory
setwd(" ")
library("boot")
x <- c(0.245, 0.247, 0.285, 0.299, 0.327, 0.347, 0
    .356,
    0.360, 0.363, 0.364, 0.398, 0.400, 0.409, 0.421,
    0.432, 0.473, 0.509, 0.529, 0.561, 0.569, 0.594,
    0.638, 0.656, 0.816, 0.853, 0.938, 1.036, 1.045)
y <- c(0, 0, 1, 1, 1, 1, 0, 1, 0, 1, 0, 1, 0, 1, 0, 1,
    1, 1, 0, 0, 1, 1, 1, 1, 1, 1, 1, 1)
xbar <- mean(x)
corr_x <- x-xbar
Nobs <- length(y)
# read posterior samples
k <- 25000   # k = posterior sample size
# Initialize matrices to hold BUGS output
alpha <- matrix(0,nrow=k,ncol=1)
beta <- matrix(0,nrow=k,ncol=1)
# read BUGS output
z <- read.table("spider.out",header=FALSE,row.names =
    NULL)
alpha[,1] <- z[1:k,2]
beta[,1] <- z[(k+1):(2*k),2]
pcor <- rep(0,Nobs)
p <- rep(0,Nobs)
for (i in 1:k){
   # parameters have index i
   for (j in 1:Nobs){
      # observations have index j
      p[j] <- inv.logit(alpha[i]+beta[i]*corr_x[j])
   }
   pred <- (p >= 0.5)
   pcor <- pcor + (y == pred)
}
pcor <- pcor/k
par(mai=c(1,1,0.75,1))
plot(y,pcor,xlab="observed
   presence/absence",ylab="P(correct)",
      pch=1,type =
          "p",ylim=c(0,1),xlim=c(0,1),xaxp=c(0,1,1))
abs <- pcor[y==0]
p_corr_abs <- mean(abs)
p_corr_abs
pres <- pcor[y==1]
p_corr_pres <- mean(pres)
p_corr_pres
```

The posterior probability of classifying each beach correctly is shown in Fig. 7.8, where the beaches are separated by whether spiders were present (1) or absent (0). It is clear from the figure that for most of the beaches without spiders, the probability

Fig. 7.8 Posterior
probability of correct
classification for spider data,
separated by whether spiders
were present (1) or absent (0)
on beach

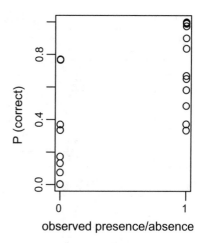

of a correct classification is below 0.5. In fact the mean probability of a correct
classification was 0.29 for beaches where spiders were absent, and 0.83 for beaches
where spiders were present. The poor performance of the model in classification is
in agreement with Figs. 7.6 and 7.7, and with the earlier metrics in Table 7.3 which
showed that the model with the covariate grain size was not appreciably better than
the model without it.

Given the above results, an immediate question arises as to whether the results
were due to an imbalance in the data (there were 9 beaches where spiders were absent
and 19 where they were present). To check this, we performed a number of splits of
the data in which we retained all the data from beaches on which spiders were absent
and randomly chose 9 beaches on which spiders were present (without replacement),
and then re-fitted the model. The results did not improve, leading us to conclude that
the model is suspect and that the imbalanced data is not the problem.[6]

7.2.2 Binomial Logistic Example

The binomial logistic model is similar to the Bernoulli logistic model, except that
in the former each observation consists of the number of successes in a set of trials,
i.e., the response variable is a proportion or percentage. For example, we might be
studying a particular species of bird and our interest might be in the percentage of
successful nests in an area, which would of course be found by dividing the number
of successful nests the total number of nests. Or we might be interested in the percent
mortality when a known number of animals are exposed to a given concentration of

---

[6]Our results also agree with the r^2 value of 0.133 reported by Suzuki et al. (2006), obtained using
non-Bayesian methods.

Table 7.4 Mortality of beetles exposed to known concentration of carbon disulphide (Bliss 1935)

Replication	Concentration	Total	Dead
1	49.06	29	2
1	52.99	30	7
1	56.91	28	9
1	60.84	27	14
1	64.76	30	23
1	68.69	31	29
1	72.61	30	29
1	76.54	29	29
2	49.06	30	4
2	52.99	30	6
2	56.91	34	9
2	60.84	29	14
2	64.76	33	29
2	68.69	28	24
2	72.61	32	32
2	76.54	31	31

a chemical. In any case, the binomial logistic model is useful when the data model is a binomial model, rather than a Bernoulli model as in Sect. 7.2.1.

Consider the example "Beetles: logistic, probit and extreme value models" in Examples, Volume II of **OpenBUGS**. The example presents data from Bliss (1935) on the mortality of the confused flour beetle (*Tribolium confusum*) in response to known concentrations of carbon disulphide. There were two replications of the study and the data are presented in Table 7.4.[7] Assuming the objective of the study was to assess the effect of carbon disulphide concentration on the probability of mortality, the following model suggests itself:

$$y_{ij} \sim \mathbf{Bi}(n_{ij}, p_{ij}), \qquad (7.2.8)$$

$$logit(p_{ij}) = \alpha_i + \beta_i x_{ij}, \qquad (7.2.9)$$

$$\alpha_i \sim \mathbf{N}(\alpha_0, \sigma_{\alpha_0}^2), \quad \beta_i \sim \mathbf{N}(\beta_0, \sigma_{\beta_0}^2) \qquad (7.2.10)$$

$$\alpha_0 \sim \mathbf{N}(0, 1.0 \times 10^6), \quad \beta_0 \sim \mathbf{N}(0, 1.0 \times 10^6) \qquad (7.2.11)$$

$$\sigma_{\alpha_0} \sim \mathbf{Unif}(0, 100), \quad \sigma_{\beta_0} \sim \mathbf{Unif}(0, 100) \qquad (7.2.12)$$

$$i = 1, 2, \quad j = 1, \ldots, n_i, \qquad (7.2.13)$$

where n_{ij} is the number of insects exposed on trial j, replication i, and n_i is the number of trials for replication i.

In (7.2.8)–(7.2.13), we postulate a hierarchical model in which replication-specific intercepts and slopes in the logit link function are specified. The intercepts and slopes

---

[7]In the **OpenBUGS** example, the two replications were combined.

are modeled as realizations from Normal priors. Finally, the parameters of the priors are modeled as realizations from vague hyperpriors. Once again, we do not include an error term in the logistic regression (7.2.9) since the probabilities of success are not observations. The **OpenBUGS** code to fit this model is presented in Box 7.8.

Code box 7.8 OpenBUGS code for fitting hierarchical binomial-logistic model to beetle data.

```
Model
{
for (i in 1:n) {
  logc[i] ← log(conc[i])
  dead[i]~ dbin(p[i],total[i])
  logit(p[i]) ← alpha[rep[i]] +
                beta[rep[i]]*(logc[i]-mean_logc)
  }
for (j in 1:nrep) {
  alpha[j] ~ dnorm(alpha0 ,tau_alpha0)
  beta[j] ~ dnorm(beta0, tau_beta0)
  }
alpha0 ~ dnorm(0, 1.0E-6)
beta0 ~ dnorm(0, 1.0E-6)
std_alpha0 ~ dunif(0,100)
tau_alpha0 ← 1/pow(std_alpha0,2)
std_beta0 ~ dunif(0,100)
tau_beta0 ← 1/pow(std_beta0,2)
alpha_dif ← alpha[1]-alpha[2]
beta_dif ← beta[1]-beta\citet{ch7CarlinLouis09}
}
DATA
list( n=16, nrep=2, mean_logc = 4.129509,
rep = c(1, 1, 1, 1, 1, 1, 1, 1, 2, 2, 2, 2, 2, 2, 2,
    2),
conc = c(49.06, 52.99, 56.91, 60.84, 64.76, 68.69,
      72.61, 76.54, 49.06, 52.99, 56.91, 60.84, 64.76,
      68.69, 72.61, 76.54),
total = c(29, 30, 28, 27, 30, 31, 30, 29, 30, 30, 34,
      29, 33, 28, 32, 31),
dead = c(2, 7, 9, 14, 23, 29, 29, 29, 4, 6, 9,
    14,
      29, 24, 32, 31) )
INITS
list( alpha=c(1,1), beta = c(0,0), alpha0=1, beta0=0,
      std_alpha0=1, std_beta0=1 )
```

In Box 7.8, y_{ij} is the number of dead beetles observed and x_{ij} is the concentration of carbon disulphide for replication i and observation j. The number of Bernoulli trials at each concentration and replication is the total number of beetles exposed at that concentration and replication. We follow Bliss (1935) and assume that the covariate is the log of the concentration of carbon disulphide. As usual, the normal distributions in the **OpenBUGS** code are parameterized by their mean and precision. Also, as in Sect. 7.2.1 we center the covariate. We *could* have used the **OpenBUGS** function mean() to compute the mean log-concentration, but it was just as easy to compute it a-priori and input it as data; either method is fine. Note also that the carbon disulphide concentrations are the same for each replication, so we only need to compute one mean for the covariate. If the concentrations had differed between the two replications, we would have needed to compute a mean log-concentration for each.

The code in Box 7.8 is for a hierarchical model in which replication-specific intercepts and slopes were specified. We also fitted the following, non-hierarchical model, in which the intercept and slope did not vary by replication:

$$y_{ij} \sim \mathbf{Bi}(n_{ij}, p_{ij}), \tag{7.2.14}$$

$$logit(p_{ij}) = \alpha + \beta x_{ij}, \tag{7.2.15}$$

$$\alpha \sim \mathbf{N}(0, 1.0 \times 10^6), \quad \beta \sim \mathbf{N}(0, 1.0 \times 10^6) \tag{7.2.16}$$

$$i = 1, 2, \quad j = 1, \ldots, n_i. \tag{7.2.17}$$

The OpenBUGS code to fit the model described in Eqs. 7.2.14–7.2.17 is presented in Box 7.9.

Code box 7.9 OpenBUGS code for fitting non-hierarchical binomial-logistic model to beetle data.

```
Model
{
for (i in 1:n) {
  logc[i] ← log(conc[i])
  dead[i]∼ dbin(p[i],total[i])
  logit(p[i]) ← alpha + beta*(logc[i]-mean_logc)
    }
alpha ∼ dnorm(0, 1.0E-6)
beta ∼ dnorm(0, 1.0E-6)
}
DATA
list( n=16, mean_logc = 4.129509,
conc = c(49.06, 52.99, 56.91, 60.84, 64.76, 68.69,
    72.61, 76.54, 49.06, 52.99, 56.91, 60.84, 64.76,
       68.69, 72.61, 76.54),
total = c(29, 30, 28, 27, 30, 31, 30, 29, 30, 30, 34,
    29, 33, 28, 32, 31),
dead = c(2, 7, 9, 14, 23, 29, 29, 29, 4, 6, 9,
   14,
    29, 24, 32, 31) )
INITS
list( alpha=1, beta = 0 )
```

The models in Boxes 7.8 and 7.9 both ran quickly. We ran each for 50,000 iterations and discarded the initial 25,000 iterations. DIC values were obtained from OpenBUGS; LOO and WAIC were obtained using the R code in Box 7.10.

Code box 7.10 R code for computing WAIC and LOO for OpenBUGS models in Boxes 7.8 and 7.9.

```
rm(list=ls())
# Analysis of Beetle data set: WAIC and LOO for
   hierarchical
```

```
# and non-heirarchical models
#  Be sure to set working directory!!
setwd(" ")
library("nlme")
library("loo")
library("boot")
#  Data:
series <- c(rep(1,8),rep(2,8))
conc <- c(49.06, 52.99, 56.91, 60.84, 64.76, 68.69, 72
    .61, 76.54)
conc <- c(conc,conc)
logcon <- log(conc)
corr_conc <- logcon - mean(logcon)
pctkill <- c( 6.9, 23.3, 32.9, 51.9, 76.7, 93.6,  96
    .7, 100.0,
                13.3, 20.0, 26.5, 48.3, 87.9, 85.7, 100
                  .0, 100.0)
total <- c(29, 30, 28, 27, 30, 31, 30, 29,
            30, 30, 34, 29, 33, 28, 32, 31)
dead <- round(pctkill*total/100)
beetle <- cbind(series,conc,total,dead)
Nobs <- length(dead)
k <- 25000  # k = posterior sample size
#  Hierarchical model
# Initialize matrices to hold BUGS output
alpha <- matrix(0,nrow=k,ncol=2)
beta <- matrix(0,nrow=k,ncol=2)
# read BUGS output from hierarchical model
y <-
    read.table("beetle_hier.out",header=FALSE,row.names
    = NULL)
alpha[,1] <- y[1:k,2]
alpha[,2] <- y[(k+1):(2*k),2]
beta[,1] <- y[((4*k)+1):(5*k),2]
beta[,2] <- y[((5*k+1)):(6*k),2]
# initialize log-likelihood matrices
log_lik <- matrix(0,nrow=k,ncol=Nobs)
# Compute log-likelihood for each tree and each set of
# parameters in joint posterior sample
for (i in 1:k){
        # parameters have index i
        for (j in 1:Nobs){
                # observations have index j
                p <- inv.logit(alpha[i,series[j]]+
                    beta[i,series[j]]*corr_conc[j])
                log_lik[i,j] <-
                    dbinom(dead[j],size=total[j],
                                    prob=p,log=TRUE)
        }
}
# Get LOO and WAIC from loo
LOOh <- loo(log_lik)
WAICh <- waic(log_lik)
```

```
LOOh
WAICh
#######################################################
# Non-hierarchical model
# Initialize matrices to hold BUGS output
alpha <- matrix(0,nrow=k,ncol=1)
beta <- matrix(0,nrow=k,ncol=1)
# read BUGS output from non-hierarchical model
y <- read.table("beetle_non_hier.out",header=FALSE,
     row.names = NULL)
alpha[,1] <- y[1:k,2]
beta[,1]  <- y[(k+1):(2*k),2]
# initialize log-likelihood matrices
log_lik <- matrix(0,nrow=k,ncol=Nobs)
# Compute log-likelihood for each tree and each set of
# parameters in joint posterior sample
#
for (i in 1:k){
  # parameters have index i
  for (j in 1:Nobs){
    # observations have index j
    p <- inv.logit(alpha[i]+beta[i]*corr_conc[j])
    log_lik[i,j] <- dbinom(dead[j],size=total[j],
                           prob=p,log=TRUE)
  }
}
# Get LOO and WAIC from loo
LOOnh <- loo(log_lik)
WAICnh <- waic(log_lik)
LOOnh
WAICnh
```

DIC, WAIC, and LOO are presented for both the hierarchical and non-hierarchical models in Table 7.5. The values in the table strongly favor the non-hierarchical model.

Boxplots and density strips for the 16 p_i values obtained with the non-hierarchical logistic regression are presented in Figs. 7.9 and 7.10, respectively. Unlike the spider data of the previous section, the binomial logistic model yields a good fit to the beetle data, as evidenced by the relatively tight bounds around each p_i.

Table 7.5 DIC, WAIC, and LOO for hierarchical and non-hierarchical beetle models

Model	DIC	WAIC	LOO
Hierarchical	65.3	66.3	66.7
Non-hierarchical	60.2	62.4	62.5

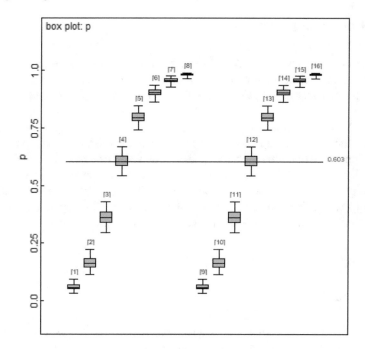

Fig. 7.9 Box plots of p_i for non-hierarchical logistic regression fitted to beetle data

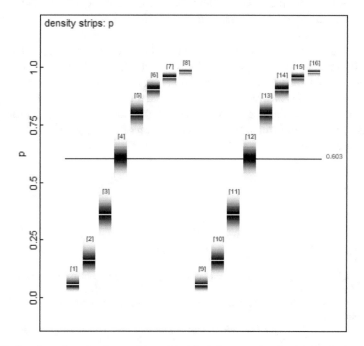

Fig. 7.10 Density strips of p_i for non-hierarchical logistic regression fitted to beetle data

7.3 Concluding Remarks

We have only scratched the surface of GLMs in this chapter. Interested readers can pursue this subject in more depth by consulting, among many others, Carlin and Louis (2009), Clark (2007), Congdon (2006), Gelman et al. (2013), and Lunn et al. (2013).

A GLM model that is of particular interest to Ecologists studying species distribution is the zero-inflated Poisson (ZIP) model and the related hurdle model. These models postulate that observing a species is a two-step process. First there is probability that an animal is present, and second there is the probability that an animal is seen, given that it's present. The latter probability accounts for the fact that for furtive species, animals may be present but not detected. These types of models are covered extensively in Kery (2010), Kery and Schaub (2012), and Williams et al. (2002).

7.4 Exercises

1. The data set `crab_data.txt` can be found online. This is a subset of data from a study on horseshoe crabs (Brockmann 1996; Guan et al. 2017), and includes 173 observations on carapace width of female horseshoe crabs (X) and the corresponding number of satellite male crabs (Y). Use Poisson regression to determine if carapace width is a useful covariate for predicting the number of satellite males.
2. The data set `newt_data.txt` can be found online. This is a subset of data from a study on newt presence/absence in the United Kingdom (Logistic regression: model building 2019), and contains 200 observations on newt presence/absence (1=present; 0=absent), percent cover of macrophytes, number of other ponds within 1 km, and the percent shadiness of the pond. Use a Bernoulli logistic model to determine which (if any) combination of the 3 covariates is best suited to predicting newt presence/absence.

References

Bliss, C. I. (1935). The calculation of the dosage-mortality curve. *The Annals of Applied Biology*, *22*(1), 134–164.

Brockmann, H. J. (1996). Satellite male groups in horseshoe crabs *Limulus polyphemus*. *Ethology*, *102*, 1–21.

Carlin, B. P., & Louis, T. A. (2009). *Bayesian Methods for Data Analysis* (3rd ed.). Boca Raton: Chapman & Hall/CRC.

Clark, J. S. (2007). *Models for Ecological Data: An Introduction*. Princeton: Princeton University Press.

Congdon, P. (2006). *Bayesian Statistical Modeling* (2nd ed.). New York: Wiley.

Gelman, A., Carlin, J. B., Stern, H. B., Dunson, D. B., Vehtari, A., & Rubin, D. B. (2013). *Bayesian Data Analysis* (3rd ed.). New York: Chapman & Hall/CRC.

Guan, Y., Yang, X. & Wan, J. (2017). Poisson regression. https://jbhender.github.io/Stats506/F17/Projects/Poisson_Regression.html. Accessed 19 May 2020.

Kery, M. (2010). *Introduction to WinBUGS for Ecologists*. Burlington: Elsevier.

Kery, M., & Schaub, M. (2012). *Bayesian population analysis using WinBUGS*. Waltham: Elsevier.

Logistic regression: model building (2019). https://www.dataanalytics.org.uk/logistic-regression-model-building/. Accessed 19 May 2020.

Lunn, D., Jackson, C., Best, N., Thomas, A., & Spiegelhalter, D. (2013). *The BUGS Book: A Practical Introduction to Bayesian Analysis*. Boca Raton: CRC Press.

McDonald, J. H. (2014). *Handbook of Biological Statistics* (3rd ed.). Baltimore: Sparky House Publishing.

Suzuki, S., Tsurusaki, N., & Kodama, Y. (2006). Distribution of an endangered burrowing spider Lycosa ishikariana in the San'in Coast of Honshu, Japan (Araneae: Lycosidae). *Acta Arachnologica*, *55*(2), 79–86.

Williams, B. K., Nichols, J. D. J., & Conroy, M. (2002). *Analysis and Management of Animal Populations*. San Diego: Academic.

Chapter 8
Spatial Linear Models

Proliferation of spatially indexed data (i.e., variable measurements are associated with a spatial location) has spurred considerable development in statistical modeling. Key texts in this field include Cressie (1993), Cressie and Wikle (2011), Chiles and Delfiner (1999), Møller and Waagepetersen (2003), Schabenberger and Gotway (2004), Wackernagel (2003), Diggle and Ribeiro (2007), and Banerjee et al. (2014). The statistical literature acknowledges that spatial (and temporal) associations are captured most effectively using models that build dependencies in different stages or hierarchies. As illustrated in Sect. 6.2, hierarchical models are especially advantageous with data sets that have several lurking sources of uncertainty and dependence. As developed in previous sections, inference about hierarchical model components is effectively pursued using Bayesian methods, see, e.g., Gelman et al. (2013), Carlin and Louis (2009).

As we have seen, computational advances in MCMC methods and software have contributed enormously to the popularity of hierarchical models, and spatial modeling is no exception. Spatial data, and hence the kind of models used to analyze them, can be roughly placed into two categories: (1) *areal* data comprise measurements over polygons, e.g., states, counties, watersheds, or forest stands; (2) *point-referenced* data comprise measurements associated with a point or location, e.g., weather stations, trees, or perhaps resource inventory plots. Areal data are most commonly analyzed using Conditionally Autoregressive (CAR) models that are easily implemented using MCMC methods such as the Gibbs sampler. In fact, this class of models is somewhat naturally suited for the Gibbs sampler which draws samples from conditional distributions that are fully specified by the CAR model. Popularity of CAR models has increased in no small measure due to their automated implementation in the Open-BUGS software package. While we do see application of CAR models in natural resource management, point-referenced data are more common; and hence the focus of this chapter.

Here, again, our interest is in specifying a regression model that explains sources of variability in the dependent variable using a sample of dependent variable mea-

E. J. Green et al., *Introduction to Bayesian Methods in Ecology and Natural Resources*, https://doi.org/10.1007/978-3-030-60750-0_8

surements at a finite set of locations and, perhaps, a set of spatially coinciding independent variable measurements. In such settings, our interest is often in predicting the dependent variable at a set of unobserved locations within a specified domain. For our purpose, the domain is a geographic extent, e.g., delineated by a polygon, within which we might also define an observable condition for which the dependent variable exists, e.g., forestland within Michigan. When thinking about a model for spatial data, it is helpful to recall Waldo Tobler's First Law of Geography "everything is related to everything else, but near things are more related than distant things" (Tobler 1970). Tobler's statement is often true for ecological variables due to a host of local (e.g., soil characteristics, disturbance history, species composition, and genetics) and global (e.g., soil parent material and climate) factors.

If prediction is our goal then we might write down a regression model that has a set of independent variables that describe well the spatial variability of the dependent variable. If the independent variables "explain" all of the spatial variability in the dependent variable, and there are no other forms of unexplained serial correlation, then the residuals should be independent and identically distributed (*i.i.d.*) as implied by the definition of *e* in Eq. 6.0.1. However, in practice, it is rare that available independent variables explain enough variability in the dependent variable to satisfy the distributional assumption of the residual term. In such cases it is useful to add a term that jointly models the spatial dependency among all observed and unobserved measurements on the dependent variable that is not already explained by the independent variables. Ideally such a term would provide a mechanism to accommodate the First Law of Geography. Inclusion of this term in the model mean would yield *i.i.d.* residuals and improve inference by facilitating the borrowing of information from observed locations to inform prediction at unobserved locations.

8.1 Point-Referenced Spatial Models

One generic paradigm that delivers a "bona fide" joint distribution able to accommodate multiple sources of variation is that of the Bayesian hierarchical model (Berliner 1996) of the form[1]

$$[\text{process, parameters} \mid \text{data}] \propto [\text{data} \mid \text{process, parameters}] \tag{8.1.1}$$
$$\times [\text{process} \mid \text{parameters}] \times [\text{parameters}].$$

How exactly the underlying process is specified depends upon the scientific hypothesis and data attributes relevant to inference (see, e.g., Gelman et al. (2014), Banerjee et al. (2014)). In the current setting, with point-referenced datasets, where spatial locations are indexed by coordinates on a map, the "process" is modeled as

---

[1]To see where this expression comes from, let θ represent the unknown parameters and d the observed data. Then recall $p(\theta \mid d) = p(d \mid \theta)p(\theta)/p(d) \propto p(d \mid \theta)p(\theta)$, since $p(d)$ is constant after the data are observed.

a spatial random field over the domain and the observations are treated as a finite realization of this random field. For example, endowing the finite-dimensional real-izations of a random field with a probability law that is multivariate normal leads to the Gaussian process (GP). The GP is, perhaps, the most conspicuous of process specifications and offers flexibility and richness in modeling. The GP's popularity as a modeling tool is enhanced due to its extensibility to multivariate and spatial-temporal geostatistical settings, and comparatively greater theoretical tractability (Stein 1999). A zero centered GP on the variable w is denoted as $w(s) \sim GP(0, C(\cdot, \cdot; \theta))$ for $s \in D$, where s is a vector defining a spatial coordinate, e.g., longitude and latitude, D is the domain, and $C(\cdot, \cdot; \theta)$ is a positive definite covariance function that takes a set of parameters θ. The form and extent of spatial dependence between any two loca-tions is captured by the covariance function $C(s, s^*; \theta) = \text{Cov}(w(s), w(s^*))$, where s and s^* are two generic locations. The process realizations are collected into a $n \times 1$ vector, say $w = (w(s_1), w(s_2), \ldots, w(s_n))^\top$ where superscript $\top$ is the transpose operator, which follows a multivariate normal distribution $\mathbf{MVN}(\mathbf{0}, \Sigma)$ where Σ is the $n \times n$ covariance matrix with the (i, j)th element given by $C(s_i, s_j; \theta)$. Clearly, the function used for $C(s, s^*; \theta)$ must result in a symmetric and positive definite matrix Σ. Such functions are known as positive definite functions, details of which can be found in Cressie (1993), Chiles and Delfiner (1999), Banerjee et al. (2014), among others.

For univariate models, we customarily specify $C(s, s^*; \theta) = \sigma^2 \rho(s, s^*; \phi)$ where $\theta = \{\sigma^2, \phi\}$ and $\rho(\cdot; \phi)$ is a positive support correlation function with ϕ comprising one or more parameters that control the rate of correlation decay and smoothness of the process. The spatial process variance is given by σ^2, i.e., $\text{Var}(w(s)) = \sigma^2$. This covariance function yields a *stationary*, i.e., constant variance, and *isotropic*, i.e., correlation depends only on the Euclidean distance separating locations, pro-cess (see, e.g., Banerjee et al. 2014 for additional discussion and process models for non-constant variance and/or directional or functional dependence). The Mátern cor-relation function is a flexible class of correlation functions with desirable theoretical properties (Stein 1999) and is given by

$$\rho(\|s - s^*\|; \phi) = \frac{1}{2^{\nu-1}\Gamma(\nu)}(\phi\|s - s^*\|)^\nu \mathcal{K}_\nu(\|s - s^*\|; \phi); \quad \phi > 0, \ \nu > 0,$$

(8.1.2)

where $\|s - s^*\|$ is the Euclidean distance between s and s^*, $\phi = \{\phi, \nu\}$ with ϕ con-trolling the rate of correlation decay and ν controlling the process smoothness, Γ is the Gamma function, and $\mathcal{K}_\nu$ is a modified Bessel function of the third kind with order ν. Figure 8.1 illustrates the Mátern correlation function for different values of ν and ϕ. When conducting analysis, it is often useful to provide a summary of the geographic extent of correlation. However, the Mátern, as well as correlation functions commonly derived from it, never actually reaches zero correlation.[2] In such cases, we generally use a reference correlation of 0.05 and call the distance at which this value is reached

---

[2]This agrees with Tobler's First Law since it allows all observations to be correlated with each other but near ones to exhibit higher correlation than more distant ones.

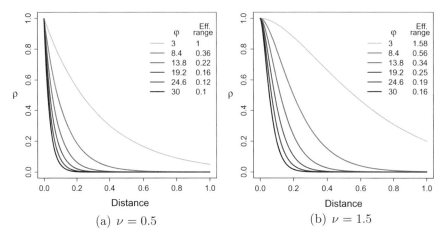

Fig. 8.1 Mátern function correlation for two values of ν and six values of ϕ. The effective spatial range (abbreviated *Eff. range* in the legends) is the distance at which the correlation equals 0.05

the *effective spatial range*, see, e.g., Fig. 8.1 for some examples. While it is theoretically ideal to estimate both ϕ and ν, it is often useful from a computational standpoint to fix ν and estimate only ϕ. For many practical problems, such a concession is reasonable given there might be little information gain in estimating both parameters. Conveniently, when $\nu = 0.5$ the Mátern correlation reduces to the familiar exponential correlation function, i.e., $\rho(||s - s^*||; \phi) = \exp(-\phi||s - s^*||)$. The effective spatial range, say d_0, for the exponential correlation function is $d_0 = -log(0.05)/\phi \approx 3\phi$ (this comes in handy when choosing the geographic support for ϕ's prior distribution).

To build some intuition about the GP, it is useful to generate and map realizations given different covariance matrix parameter values. The R code in Box 8.1 provides the essentials for generating realizations from a spatial GP. Here, we draw one realization at $n = 2000$ locations distributed randomly within a unit square domain. Process parameters are fixed at $\sigma^2 = 1$ and $\phi = 3$. This value of ϕ corresponds to an effective spatial range of 1, which is long relative to the size of the domain. In the code, we first define the rmvn function that returns realizations from a n-dimensional **MVN** given a n length mean vector mu and $n \times n$ covariance matrix Sigma. Next we define the process parameters σ^2 and ϕ (i.e., sigma.sq and phi). Then to construct the process' covariance matrix we compute the $n \times n$ Euclidean distance matrix D, which is used to form the exponential correlation matrix R. Scaling R by sigma.sq yields the desired covariance matrix. As detailed in the GP definition above, given the covariance matrix and mean vector (which we set to zeros), the process realizations w are then drawn from a **MVN**.

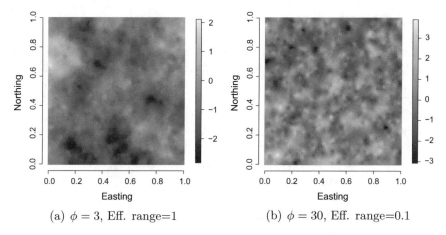

(a) $\phi = 3$, Eff. range=1   (b) $\phi = 30$, Eff. range=0.1

Fig. 8.2 Surface plots of realizations from a Gaussian process using an exponential spatial correlation function with $\sigma^2 = 1$ and ϕ equal 3 and 30 for **a** and **b**, respectively

Code box 8.1 R code to generate realizations from a spatial GP.

```
rmvn <- function(mu, Sigma){
    mu + t(chol(Sigma))%*%rnorm(length(mu))
}
n <- 2000
coords <- cbind(runif(n,0,1), runif(n,0,1))
sigma.sq <- 1
phi <- 3
D   <- as.matrix(dist(coords))
R   <- exp(-phi*D )
Sigma <- sigma.sq*R
w <- rmvn(rep(0,n), Sigma)
```

Surface plots of w using $\phi = 3$ and $\phi = 30$ are given in Fig. 8.2a, b, respectively. Visual inspection of these two figures provides a sense of what long-range (e.g., $\phi = 3$) versus short-range (e.g., $\phi = 30$) looks like.

In a regression context, the GP is included as a random effect in the model and data are used to inform process parameter estimates. We begin with a simple space-varying intercept model

$$y(s_i) = x(s_i)^\top \beta + w(s_i) + e(s_i), \quad i = 1, 2, \ldots, n, \qquad (8.1.3)$$

where $y(s)$ is a Gaussian dependent variable, $x(s)$ is a $p \times 1$ vector that includes an intercept (value of one) and spatially referenced independent variables, β is a $p \times 1$ vector of regression coefficients,[3] $w(s)$ is the zero centered GP random effect, and $e(s)$ is the i.i.d residual that follows $N(0, \tau^2)$. In the geostatistical literature, the

---

[3] $x(s)^\top \beta$ is matrix multiplication notation involving the transpose of $x(s)$ and β which, in this case, is equivalent to the sum of the element-wise products $\sum_{j=1}^{p} x_j(s)\beta_j$.

residual variance parameter τ^2 is often referred to as the *nugget* variance and can be interpreted as measurement error.

The Bayesian specification is completed by assigning prior distributions to the model parameters. A common setup is $\boldsymbol{\beta}$ following a p-dimensional $\mathbf{MVN}(\boldsymbol{\mu}_\beta, \boldsymbol{\Sigma}_\beta)$ or flat prior, inverse-Gamma $\mathbf{IG}(\cdot, \cdot)$ for the variance parameters, i.e., $\sigma^2 \sim \mathbf{IG}(a_\sigma, b_\sigma)$ and $\tau^2 \sim \mathbf{IG}(a_\tau, b_\tau)$, and a uniform prior for $\phi \sim \mathbf{Unif}(a_\phi, b_\phi)$ (and v if estimated). The parameters' joint posterior distribution $p(\boldsymbol{\beta}, \boldsymbol{w}, \sigma^2, \tau^2, \phi \mid \boldsymbol{y})$, where $\boldsymbol{y} = (y(\boldsymbol{s}_1), y(\boldsymbol{s}_2), \ldots, y(\boldsymbol{s}_n))^\top$, is proportional to

$$
\begin{aligned}
&\mathbf{IG}(\tau^2 \mid a_\tau, b_\tau) \times \mathbf{IG}(\sigma^2 \mid a_\sigma, b_\sigma) \times \mathbf{Unif}(\phi \mid a_\phi, b_\phi) \qquad (8.1.4) \\
&\times \mathbf{MVN}(\boldsymbol{\beta} \mid \boldsymbol{\mu}_\beta, \boldsymbol{\Sigma}_\beta) \times \mathbf{MVN}(\boldsymbol{w} \mid \boldsymbol{\mu}_w, \boldsymbol{\Sigma}_w) \\
&\times \prod_{i=1}^{n} \mathbf{N}(y(\boldsymbol{s}_i) \mid \boldsymbol{x}(\boldsymbol{s}_i)^\top \boldsymbol{\beta} + w(\boldsymbol{s}_i), \tau^2).
\end{aligned}
$$

Explicitly updating the vector of random effects often leads to an inefficient MCMC, i.e., high within chain autocorrelation and slow mixing among chains (see, e.g., Gelman et al. (2013) for discussion on desirable MCMC qualities). Therefore, in practice, we often integrate out $\boldsymbol{w}$ and call the resulting model the *collapsed* form of the *latent* model (8.1.4). The remaining parameters' joint posterior distribution $p(\boldsymbol{\beta}, \sigma^2, \tau^2, \phi \mid \boldsymbol{y})$ for the collapsed model is proportional to

$$
\begin{aligned}
&\mathbf{IG}(\tau^2 \mid a_\tau, b_\tau) \times \mathbf{IG}(\sigma^2 \mid a_\sigma, b_\sigma) \times \mathbf{Unif}(\phi \mid a_\phi, b_\phi) \qquad (8.1.5) \\
&\times \mathbf{MVN}(\boldsymbol{\beta} \mid \boldsymbol{\mu}_\beta, \boldsymbol{\Sigma}_\beta) \times \mathbf{MVN}(\boldsymbol{y} \mid \boldsymbol{X}\boldsymbol{\beta}, \sigma^2 \boldsymbol{R}(\phi) + \tau^2 \boldsymbol{I}),
\end{aligned}
$$

where $\boldsymbol{X}$ is the $n \times p$ design matrix with the ith row as $\boldsymbol{x}(\boldsymbol{s}_i)^\top$ and $\boldsymbol{I}$ is the $n \times n$ identity matrix. While the collapsed model does greatly improve MCMC sampler efficiency, it does not provide immediate access to the posterior samples of $\boldsymbol{w}$. From an analysis perspective, these spatial random effects are often useful because they portray spatial variability in the dependent variable that is not accounted for by independent variables. Fortunately, we can recover posterior samples of $\boldsymbol{w}$ in a posterior predictive manner given samples from $\boldsymbol{\beta}$, σ^2, τ^2, and ϕ; see Finley et al. (2015) for details.

8.1.1 Space-Varying Coefficient Models

Model (8.1.3) can be generalized in a straightforward way to accommodate a very rich set of spatial models. One extension called a spatially varying coefficient (SVC) model posits that independent variables might have a space-varying impact on the dependent variable (Gelfand et al. 2003; Finley 2011). In such settings, regression coefficient specific GP's capture a global mean and continuous, smoothly varying, location specific adjustment. The SVC model can be written as

$$y(s_i) = (\beta_1 + \delta_1 w_1(s_i)) + \sum_{j=2}^{p} x_j(s_i) \{\beta_j + \delta_j w_j(s)\} + e(s_i), \ i = 1, 2, \ldots, n,$$

(8.1.6)

where $x_j(s)$, for each $j = 2, \ldots, p$, is the known value of a independent variable at location s, β_j is the regression coefficient corresponding to $x_j(s)$, β_1 is an intercept, and $e(s)$ was previously defined for Model (8.1.3). The quantities $w_1(s)$ and $w_j(s)$ are spatial random effects corresponding to the intercept and independent variables, respectively, thereby yielding a spatially varying regression model. To further accommodate the possibility that not all the independent variables will have spatially varying impact on the dependent variable, the δ's in (8.1.6) are binary indicators assuming the value 1 if the associated variable has a spatially varying regression coefficient and 0 otherwise. For later convenience, when the respective $\delta = 1$ we define $\tilde{\beta}_1(s) = \beta_1 + \delta_1 w_1(s)$ and $\tilde{\beta}_j(s) = \beta_j + \delta_j w_j(s)$ as the space-varying regression coefficients.

Consider the Bayesian hierarchical model built from (8.1.6),

$$p(\tau, \boldsymbol{\theta}) \times N(\boldsymbol{\beta} \mid \boldsymbol{\mu}_\beta, \boldsymbol{\Sigma}_\beta) \times N(\boldsymbol{w} \mid \boldsymbol{0}, \boldsymbol{K}) \times N(\boldsymbol{y} \mid \boldsymbol{X}\boldsymbol{\beta} + \boldsymbol{Z}\boldsymbol{w}, \tau^2 \boldsymbol{I}), \quad (8.1.7)$$

where $\boldsymbol{y}$ and $\boldsymbol{X}$ were defined previously, $\boldsymbol{Z}$ is a vector diagonal matrix of dimension $n \times rn$, where $r = \sum_{j=1}^{p} \delta_j$, and is constructed using precisely those columns of $\boldsymbol{X}$ that have their respective δ's equal to 1, such that row 1 and columns 1 through r hold the row 1 and $\delta = 1$ columns of $\boldsymbol{X}$, row 2 and columns $r + 1$ through $2r$ hold row 2 and $\delta = 1$ columns of $\boldsymbol{X}$, and so forth until row n and columns $nr - r$ through nr hold row n and $\delta = 1$ columns of $\boldsymbol{X}$. All elements other than the row vectors along the diagonal of $\boldsymbol{Z}$ are zero. Then, let $\boldsymbol{w}$ be the $nr \times 1$ vector obtained by stacking up $\boldsymbol{w}(s_i)$'s, where each $\boldsymbol{w}(s_i)$ is an $r \times 1$ vector with jth entry $w_j(s_i)$, $j = 1, 2, \ldots, r$ and $i = 1, 2, \ldots, n$. In this way, $\boldsymbol{w}(s_i)$ is viewed as location s_i's specific set of regression coefficients.

To account for spatial correlation and covariance among regression coefficients, we treat $\boldsymbol{w}(s)$ as a zero centered multivariate Gaussian process (see, e.g., Banerjee et al. 2014) with an $nr \times nr$ spatial covariance matrix matrix constructed as a block matrix with (i, j)th block obtained from the $r \times r$ cross-covariance matrix $\boldsymbol{K}(s_i, s_j; \boldsymbol{\theta})$. Some further specifications are in order. The cross-covariance matrix $\boldsymbol{K}(s, s^*; \boldsymbol{\theta})$ is often specified using a Linear Model of Coregionalization (LMC; Gelfand et al. 2003). In this way $\boldsymbol{K}(s, s^* \boldsymbol{\theta}) = \boldsymbol{A}\Gamma(s, s^*)\boldsymbol{A}^\top$, where $\boldsymbol{A}$ is an $r \times r$ lower triangular matrix and $\Gamma(s, s^*)$ is a diagonal matrix with the jth diagonal element $\rho_j(s, s^*; \boldsymbol{\phi}_j)$ being a spatial correlation function with parameters specific to $w_j(s)$. Here, the vector of covariance parameters $\boldsymbol{\theta}$ corresponds to $\{\boldsymbol{A}, \{\boldsymbol{\phi}_j\}_{j=1}^r\}$ where each $\boldsymbol{\phi}_j$ may itself be a collection of parameters in the spatial correlation function. For example, with the Matérn covariance function each $\boldsymbol{\phi}_j$ comprises a spatial decay parameter and a smoothness parameter. Note, ρ_j need not be the same correlation function for each j.

The covariance structure for $\boldsymbol{w}(s)$ within any location s, i.e., var$\{\boldsymbol{w}(s)\}$, equals $\boldsymbol{A}\boldsymbol{A}^\top$ and is the covariance among the spatially-varying regression coefficients. From a modeling perspective, specification of the cross-covariance function in this way is convenient because we need only to estimate the lower triangular elements in A, with the restriction that the diagonal elements are positive; then by construction we are ensured a positive definite covariance matrix (see Gelfand et al. 2003; Banerjee et al. 2014 for more details). For this model, a reasonable prior specification for the variance and covariance parameters is

$$p(\tau^2, \boldsymbol{\theta}) = \mathbf{IG}(\tau^2 \,|\, a_\tau, b_\tau) \times \mathbf{IW}(A A^\top \,|\, r_a, S_a) \times \prod_{j=1}^{r} \mathbf{Unif}(\boldsymbol{\phi}_j) \,, \qquad (8.1.8)$$

where **IW** is inverse-Wishart, and **IG** and **Unif** were previously defined (the Matérn smoothness parameter can also be added if desired). In practice, we might not be interested in estimating the covariance among the processes, or do not want to assume a stationary cross-covariance, in which case we might specify $A = \mathrm{diag}(\sigma_1, \ldots, \sigma_r)$, so that $K(s, s^*; \boldsymbol{\theta})$ is diagonal with entries $\sigma_j^2 \rho_j(s, s^*)$, in which case we assume $IG(\sigma_j^2 \,|\, a_\sigma, b_\sigma)$ for $j = 1, 2, \ldots, r$.

8.1.2   Software

There are several challenges to implementing GP-based spatial regression models. First, expensive matrix computations are required that can become prohibitive with large datasets. Second, software in the BUGS paradigm are not well suited to fitting the collapsed model forms, hence the less efficient (i.e., slower chain convergence and mixing) latent forms are implemented. Third, highly specialized software is needed to accommodate even routine data characteristics such as: space-varying relationships between dependent and independent variables, e.g., accommodated through the SVC model in Sect. 8.1.1; multivariate data sets with several spatially correlated dependent variables; spatio-temporal data have correlation in both space and time; multivariate data with missing observations, referred to as spatially misaligned data, and; when dependent variables are not Gaussian. Increasingly, however, specialized software designed to overcome computational challenges and accommodate complex process specifications is emerging. A recent read of the "Analysis of Spatial Data" CRAN Task View (Bivand 2019) yielded ~46 packages listed for geostatistical analysis—and this is not an exhaustive accounting of packages available for such analyses. Several notable packages that implement variations on model (8.1.3) include **geoR** (Ribeiro and Diggle 2018), **geoRglm** (Christensen and Ribeiro 2017), **spTimer** (Bakar and Sahu 2018), **spBayes** (Finley and Banerjee 2019), and **spNNGP** (Finley et al. 2020), that help automate Bayesian methods for point-referenced data and diagnose convergence. Here we focus on the **spBayes** package that provides a suite of univariate and multivariate (with misalignment) regression models for both Gaussian and non-Gaussian outcomes that are spatially indexed, see details in Finley et al. (2007), Finley et al. (2015).

8.1.3 Tree Height-Diameter Data

To build a bit more intuition about spatial GP models, we analyze a forestry data set that comprises height (HT) and diameter at breast height (DBH; measured 1.37 m from the ground) for 2391 mature trees measured on a 200×200 m portion of an uneven-aged softwood stand located near Sault Ste. Marie, Ontario (Ek 1969). The major tree species are balsam fir (*Abies balsamea*) and black spruce (*Picea mariana*). Location, DBH, and TH were recorded for each tree. See Guo et al. (2008) for a full description of these data. Our analysis objective is to build a model that predicts HT from DBH. Here, the candidate models use $DBH^{0.5}$ to explain the variability in the outcome variable HT.

The relationship between tree DBH and TH is influenced by several individual and environmental factors. Individual factors include age, species, and genetics, whereas, environmental factors include quality of soil, quantity of water and light, and competition for these resources. These factors can cause local spatial dependence in TH and varying relationships between DBH and TH across the domain. For example a cohort of trees of the same species, age, and parentage and hence similar DBH and TH relationship might be established in gaps created by some form of disturbance, e.g., fire, wind, or insect infestation. An example of an environmental factor influencing tree growth characteristics is proximate similarities in soil productivity due to parent material or disturbance history. Given these unobserved covariates it is reasonable to allow the coefficient associated with DBH to vary spatially. Further, we might expect additional spatial dependence among the model residuals, which can be accommodated by a space-varying intercept.

We begin with a bit of exploratory data analysis (EDA). Figure 8.3a shows the stem map (i.e., spatial location of each tree in the stand) with color indicating the given tree's height. Even if you squint, it is hard to detect much of a spatial pattern in these data; however, recall, given the posited model, we are actually interested in modeling residual spatial patterns, i.e., after accounting for $DBH^{0.5}$, which we will take a look at soon. The scatter plot of HT versus $DBH^{0.5}$, Fig. 8.3b, shows that linear regression is a reasonable predictive model. Note, without the square root transformation of the DBH, the HT versus DBH relationship is fairly non-linear. A more thorough analysis might further assess the influence of a few of the points well outside the scatter around the one-to-one line; however, given this is simply for illustration, we will not consider any data cleaning or further transformations.

It might be the case that available predictor variables account for all spatial patters in the response—hence no serial correlation among model residuals—in which case there is no need to pursue a spatial model. There are several formal and informal EDA approaches for assessing presence of spatial patterns among model residuals, see Banerjee et al. (2014), Schabenberger and Gotway (2004). Most approaches begin with fitting a non-spatial regression then inspecting the residuals ($y - \hat{y}$, where $\hat{y} = X\beta$). A simple, but informative, EDA is to map model residuals with observed data location symbol size or color reflecting the corresponding residual value. Such a map is given in Fig. 8.4a using residuals from the regression

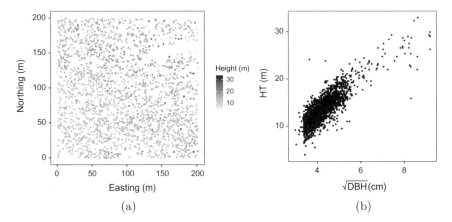

Fig. 8.3 a location of trees with color corresponding to tree height (HT). **b** TH versus square root transformed diameter at breast height (DBH)

$$\text{TH}(s_i) = \beta_1 + \beta_{DBH}\text{DBH}^{0.5}(s_i) + e(s_i), \tag{8.1.9}$$

for $i = 1, 2, \ldots, n$, where n is the 1793 locations used to estimate the model parameters (we're withholding 25% of the observations for subsequent model comparison).

In addition to mapping the residuals, an empirical semi-variogram constructed using model residuals is a valuable tool for assessing presence of spatial correlation and to inform selection of prior distributions along with associated hyperparameters. We refer the reader to Cressie (1993), Banerjee et al. (2014), Diggle and Ribeiro (2007), and similar texts that focus on spatial modeling for background reading on semi-variogram construction. These texts layout the connection between the semi-variogram model and spatial regression model (8.1.3). Referring to Fig. 8.4b, which was constructed using the `variog` and `varofit` functions in the **geoR** package, the semi-variogram's *nugget* (i.e., the lower horizontal line) is an estimate of τ^2, *partial sill* (difference between the upper and lower horizontal lines) is an estimate of σ^2, and the *range* (vertical line) is an estimate of the effective spatial range (i.e., $\sim 3/\phi$, assuming an exponential variogram model and covariance function).

Both Fig. 8.4a and Fig. 8.4b suggest there is evidence of residual spatial correlation; hence, exploring some spatial models is warranted. Here we consider both a space-varying intercept (SVI) and space-varying coefficients (SVC) model that allows both the intercept and impact of $\text{DBH}^{0.5}$ to vary across the stand. For generic location s, the posited SVI model is

$$\text{TH}(s) = \tilde{\beta}_1(s) + \beta_{DBH}\text{DBH}^{0.5}(s) + e(s) \tag{8.1.10}$$

and SVC model is

$$\text{TH}(s) = \tilde{\beta}_1(s) + \tilde{\beta}_{DBH}(s)\text{DBH}^{0.5}(s) + e(s), \tag{8.1.11}$$

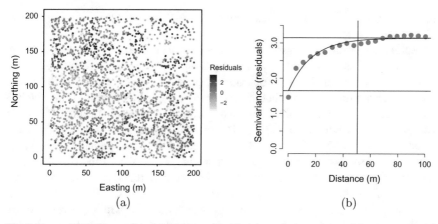

Fig. 8.4 **a** residuals from a linear regression of tree height as the response and the square root of diameter at breast height as the predictor. **b** Semi-variogram of linear regression model residuals shown in Fig. 8.4a. Curved line corresponds to a non-linear regression fit of an exponential spatial covariance function, vertical line corresponds to the estimated effective spatial range, the lower horizontal line corresponds to the estimated measurement error τ^2 and the difference between the upper and lower horizontal lines corresponds to the estimated spatial variance σ^2

where, recall, $\tilde{\beta}_1(s) = \beta_1 + w_1(s)$ and $\tilde{\beta}_{\text{DBH}}(s) = \beta_{\text{DBH}} + w_{\text{DBH}}(s)$. The three candidate models, i.e., non-spatial, SVI, and SVC, will be assessed using the fit metrics described in Sect. 5.5 as well as out-of-sample prediction performance based on a subset of 598 observations (i.e. 25% holdout set).

The subsequent analysis uses functions in the **spBayes** package to fit the three candidate models. The `bayesLMRef` function is used to fit the non-spatial model (8.1.9). This function assumes Jeffreys' priors on the model parameters and provides posterior inference via an exact sampling algorithm. The spatial models are fit using the `spSVC` function which implements a MCMC sampler for the collapsed forms of (8.1.10) and (8.1.11). As detailed in Finley and Banerjee (2020), an efficient MCMC algorithm is used in `spSVC` that integrates out the $\boldsymbol{\beta}$'s and $\boldsymbol{w}$'s and only draws samples for the covariance parameters $\boldsymbol{\theta}$ and τ^2. While the collapsed form of the model greatly reduces the number of parameters to be sampled (leading to a more efficient sampler), it requires the remaining parameters, i.e., $\boldsymbol{\theta}$ and τ^2, be updated using a Metropolis MCMC sampler (see, e.g., Carlin and Louis (2009) or Gelman et al. (2013) for details). In short, the Metropolis sampler is an accept-reject algorithm where a new parameter value is generated from a proposal distribution that is a function of the currently accepted parameter value. In `spSVC`, the proposal distribution is Normal with mean equal to the currently accepted parameter value that is transformed to have support on the real line, and variance set by the user. The variance, referred to as the *tuning* value, is set such that newly proposed parameter values are accepted between ∼30–50 percent of the time (acceptance in this range typically allows for adequate exploration of the parameter's posterior given a sufficient number of samples). After the sampler is complete, a post burn-in subset of $\boldsymbol{\theta}$ and τ^2 are passed to `spRecover` to draw β and w via composition sampling (see Finley et al. (2015), Banerjee et al. (2014) for details).

An excerpt of the code used to fit the SVI model is provided in Box 8.2 (the full code is available online). Arguments passed to the spSVC function define the spatial covariance function via cov.model (in this case we chose an exponential), parameter prior distributions via the priors list, parameter starting values via the starting list, and a set of values used to control parameter specific Metropolis acceptance rate via the tuning list. The spSVC argument n.samples sets the desired number of MCMC samples to be collected. Given the symbolic model statement ht $\sim$ dbh passed to spSVC, the argument svc.cols=1 indicates we want the spatial process to be placed on the model intercept (the design matrix column name could also be used, e.g., svc.cols="Intercept"). Following the model specification in Sect. 8.1, we use a Uniform prior distribution with support from 0.02 to 3 for ϕ and inverse-Gamma prior distribution for the variance parameters. Empirical semi-variogram estimates can serve as a rough guide for setting prior distribution hyperparameters and MCMC chain starting values. For example, following Fig. 8.4b, we center the inverse-Gamma prior distribution for σ^2 and τ^2 at approximately 1.5, i.e., $IG(a = 2, b = 1.5)$ where a is the shape and b is the scale. Importantly, given a shape of 2, the mean of the inverse-Gamma equals the scale and the variance of the distribution is infinite. The bounds for ϕ's Uniform prior are chosen based on the geographic extent of the domain, which, in this case, allow for an effective spatial range somewhere between 1 and 150 m. Recalling the discussion of the effective spatial range in Sect. 8.1, note that $-\log(0.05)/150 \approx 0.02$ and $-\log(0.05)/1 \approx 3$). Similarly, a starting value of $\phi = 0.06$ is equivalent to an effective spatial range of 50 m (i.e., $-\log(0.05)/0.06 \approx 50$ m).

The output from spSVC is then passed to spRecover which generates MCMC samples from β and w posteriors for post burn-in samples of θ and τ^2 (i.e., starting at sample iteration floor(0.75*n.samples)).

Code box 8.2 Abbreviated code for fitting the SVI model using the **spBayes** R package.

```
cov.model <- "exponential"
priors <- list("phi.Unif"=list(0.02, 3),
    "sigma.sq.IG"=list(2, 1.5), "tau.sq.IG"=c(2, 1.5))
starting <- list("phi"=0.06, "sigma.sq"=1.5,
    "tau.sq"=1.5)
tuning <- list("phi"=0.1, "sigma.sq"=0.01,
    "tau.sq"=0.01)
n.samples <- 5000

svi <- spSVC(ht ~ dbh, coords=coords,
    starting=starting, svc.cols=1, tuning=tuning,
    priors=priors, cov.model=cov.model,
    n.samples=n.samples)

svi <- spRecover(svi, start=floor(0.75*n.samples))
```

To invoke the more general SVC model, the value of the svc.cols argument is set to the design matrix column indexes 1:2 (or column names c("Intercept",

Table 8.1 Tree height-diameter analysis posterior summaries of median and 95% credible interval for the non-spatial and two spatial models

Parameter	Non-spatial	SVI	SVC
β_1	$-2.31\ (-2.77, -1.82)$	$-1.61\ (-2.11, -1.09)$	$-2.12\ (-2.58, -1.62)$
β_{DBH}	3.72 (3.61, 3.83)	3.56 (3.46, 3.67)	3.68 (3.55, 3.80)
σ_1^2	–	1.60 (1.28, 2.01)	0.37 (0.15, 0.60)
ϕ_1	–	0.13 (0.09, 0.17)	1.40 (0.60, 2.12)
σ_{DBH}^2	–	–	0.11 (0.09, 0.13)
ϕ_{DBH}	–	–	0.16 (0.12, 0.22)
τ^2	3.00 (2.81, 3.21)	1.45 (1.18, 1.72)	0.76 (0.57, 1.01)

`"dbh"`)) and the spatial process parameter information for DBH needs to be added to the `priors`, `starting`, and `tuning` lists.

Parameter estimates for the candidate models are given in Table 8.1. There are a few things to notice here. First, as might be expected, the residual variance τ^2 decreases with increasing model flexibility, i.e., from a posterior median of 3 for the non-spatial model to 0.76 for the SVC model. Second, SVI model estimates for the covariance parameters σ_1^2, ϕ_1, and τ^2 are on par with the rough semi-variogram point estimates (just a nice sanity check). Third, the spatial process parameter estimates for the intercept differ substantially between the SVI and SVC models. Specifically, the SVC's spatial process parameter point estimates of $\sigma_1^2 = 0.37$ and effective spatial range of 2.1 m (i.e., $\sim -\log(0.05)/\phi_1 = -\log(0.05)/1.40$) suggest a less variable and substantially shorter spatial range than the SVI's spatial process. Fourth, the non-negligible spatial process parameter estimates of σ_{DBH}^2 and ϕ_{DBH} in the SVC model suggest there is a potentially meaningful space-varying relationship between HT and DBH.

Surface plots of the SVI and SVC regression coefficients are shown in Fig. 8.5. These surfaces were created using output from the `spPredict` function which generates posterior predictive distribution samples given a new spatial location. The value assigned to each surface pixel is the median of that location's posterior predictive distribution. These surfaces can be helpful to better understand the model estimates and perhaps identify missing covariates based on the spatial patterns they reveal. A couple things to notice about these figures. First, as expected, the general spatial patterns seen in the residual map Fig. 8.4a are evident in Fig. 8.5a—the spatial random effect w in the SVI model captures spatial structure not accounted for by DBH. Second, with the exception of scale, SVC model's space-varying regression coefficient associated with DBH, i.e., $\tilde{\beta}_{DBH}$ Fig. 8.5c, is nearly identical to the SVI model's intercept Fig. 8.5a. This could mean a few things that might be worth further exploration: (i) there are indeed space-varying relationships between HT and DBH due perhaps to unobserved species, genetic, disturbance, or environmental factors; (ii) there are spatial patterns in HT and a non-linear relationship between HT and DBH, such that the pattern we're seeing is not really a function of spatial varia-

Table 8.2 Tree height-diameter analysis model fit and out-of-sample prediction diagnostics for the non-spatial and two spatial models

	Non-spatial	SVI	SVC
DIC	7061.28	6371.99	5706.00
p_D	3.03	617.06	1099.68
$WAIC$	7065.70	6426.05	5660.43
p_{waic}	7.35	539.05	786.95
$lppd$	−3525.50	−2673.97	−2043.26
RMSPE	1.75	1.51	1.49
CRPS	0.98	0.84	0.82

tion in the regression relationship but really a functional difference between HT and DBH. Given the fairly linear relationship shown in Fig. 8.3b, we can feel somewhat confident that unobserved factors with some spatial structure are influencing the relationship between HT and DBH. As seen in Fig. 8.5b and as suggested by the SVC model parameter estimates for σ_1^2 and ϕ_1 in Table 8.1, there is little spatial structure left after accounting for the dominant spatial patterns attributed to the space-varying impact of DBH.

Given our analysis goal is to find a good predictive model for HT given DBH and spatial location, we will defer the pursuit of a more meaningful ecological/environmental understanding to some other time. Rather we can turn to model fit and out-of-sample prediction performance given in Table 8.2 to select our preferred model. Here we see the model diagnostics DIC and WAIC both favor the spatial models and the SVC model in particular (despite the increased complexity which is reflected in the larger p_D and p_{waic} for DIC and WAIC, respectively). Out-of-sample predictive performance was based on prediction of a 25% holdout set, with metrics root mean squared prediction error (RMSPE) and continuous rank probability score (CRPS; Gneiting and Raftery 2007). Compared with RMSPE which only summarizes prediction accuracy, CRPS is a more robust measure of prediction skill because it favors models with both improved accuracy and precision. Like the model fit diagnostics, RMSPE and CRPS both favor the SVC model. The holdout observed tree measurements versus prediction for the three models is given in Fig. 8.6, where, again, we see that relative to the non-spatial regression, spatial models have a tighter distribution about the 1-to-1 line.

8.2   Models for Large Spatial Data

As we have seen in the previous section, the addition of spatial processes to model components offers a rich framework for prediction and exploring complex spatial phenomena. However, model fitting involves the inverse and determinant of Σ in

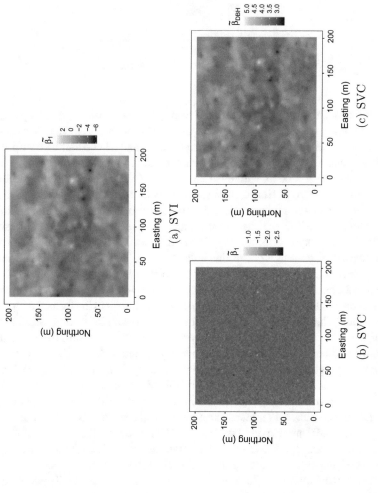

Fig. 8.5 Tree height-diameter analysis spatial models' regression coefficients

Fig. 8.6 Tree
height-diameter
out-of-sample prediction
using the 25% holdout set for
the non-spatial and two
spatial models

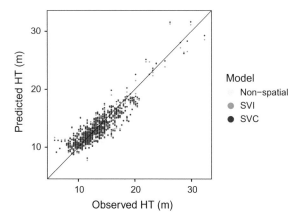

$(8.1.4)$ or $\sigma^2 \boldsymbol{R}(\phi) + \tau^2 \boldsymbol{I}$ in $(8.1.5)$, which require $\sim n^3$ floating point operations (flops) and storage of the order n^2, where again, n is the number of locations measured. Computational demands are even greater for the SVC and related multivariate process models. In a MCMC setting, these cubic order operations must occur in each iteration, which makes fitting models with even a modest n (e.g., a few thousand locations) computationally onerous. Hence, fitting such models in practice is often not practical.

Motivated by this computational hurdle, there have been many recent methodological developments within the large spatial data literature that aim to deliver massively scalable spatial processes. Sun et al. (2011), Banerjee (2017) provide background and discussion of current work in this area. A recent contribution by Heaton et al. (2019) is particularly useful as it provides an overview of modeling approaches for large spatial data that are under active software development, and a comparison of these approaches based on the analysis of a common dataset in the form of a "friendly competition." The comparison presented by Heaton et al. (2019) considered covariance tapering via the **spam** package (Furrer and Sain 2010; Furrer 2016), gapfilling via **gapfill** (Gerber 2017), metakriging (Guhaniyogi and Banerjee 2018), spatial partitioning (Sang et al. 2011; Barbian and Assunção 2017), fixed rank kriging via **FRK** (Cressie and Johannesson 2008; Zammit-Mangion and Cressie 2017), multiresolution approximation (Katzfuss 2017), stochastic partial differential equations via **INLA** (Rue et al. 2017), lattice kriging via **LatticeKrig** (Nychka et al. 2015), local approximate Gaussian processes via **laGP** (Gramacy and Apley 2015; Gramacy 2016), reduced rank predictive processes (Banerjee et al. 2008; Finley et al. 2009) via **spBayes** (Finley et al. 2015), and Nearest Neighbor Gaussian Processes (NNGP Datta et al. 2016; Finley et al. 2019 via **spNNGP** (Finley et al. 2020). The comparison was based on out-of-sample predictive performance and, to a lesser extent, computing time for a moderately sized simulated and real dataset comprising 105,569 observations. More recently, Risser and Turek (2019) developed the **BayesNSGP** package for nonstationary Gaussian process modeling with options to use NNGPs for large data settings. In a frequentist setup, fast maximum likelihood-based parameter estimation and predictions using nearest neighbor approximations to the Gaussian Process likelihood are available in the **GpGp** (Guinness 2018) and **BRISC** (Saha and Datta 2018b) packages on CRAN. The latter also offers inference on the spatial covariance parameters using a fast spatial bootstrap (Saha and Datta 2018a). While most of the software noted above exploit sparsity in the spatial covariance or precision matrix, or pursue a low-rank approximation, the **ExaGeoStat** package (Abdulah et al. 2018) tackles decomposition of the full dense spatial covariance matrix head-on using high performance linear algebra libraries associated with various leading edge parallel architectures. These and related work provide methods and associated software to facilitate fitting a variety of large spatial data modeling needs.

8.3 Exercises

1. Extend the code in Box 8.1 to simulate $n = 2000$ values from the latent spatial regression model defined in (8.1.3). You will need to define your own $n \times p$ design matrix X (set the first column to all ones, with subsequent $p - 1$ columns generated from `rnorm` or by some other random number generator), corresponding vector of p regression coefficients β, and residual variance parameter τ^2. You will then draw the realization of $y = (y(s_1), y(s_2), \ldots, y(s_n))^\top$ using `rnorm(n, mu, sd)`. Note, your own $X\beta + w$ will define the $n \times 1$ mean vector `mu` (which in R speak is `X%*%beta + w` assuming X is a matrix) and `sd` will equal τ. Alternatively, if you want to avoid matrix operations, you can set up a `for` loop over `rnorm(1, mu, sd)` with the now scalar `mu` equal to `sum(X[s,]*beta) + w[s]`, assuming s is incrementing index, i.e., `for(s in 1:n)`.

2. Following from the first exercise, generate surface plots of both w and y. This can be done in several ways in R. Perhaps the simplest is to use the `quilt.plot` function in the **fields** package, e.g., `quilt.plot(x=coords, y=w)` to map the spatial random effects. Alternatively, if you want an estimate of the complete surface, the scattered values of w and y can be passed through an interpolator then plotted using the `image` or **fields** `image.plot` function. For example, the `mba.surf` function in the **MBA** package provides fast interpolators for scattered data, such that the output from `mba.surf(cbind(coords, w), no.X=100, no.Y=100)$xyz.est` can be passed directly to `image` or `image.plot`. Change your values of ϕ, σ^2, and τ^2 to see how they effect your surface plots of w and y. Do the results follow your intuition?

3. Using the `coords`, X, and y data you generated in the first exercise, fit a spatial regression model using the `spLM` or `spSVC` function in the `spBayes` package. Do the resulting posterior distributions of σ^2, ϕ, and τ^2 capture the parameter values you used to generate the data (they should)? You should then be able to pass a model object from `spLM` or `spSVC` to `spRecover` to generate posterior samples for β and w, these too should capture well the "true" β and w used to generate y.

4. Write an **OpenBUGS** program for (8.1.3) using your simulated data and see if the resulting posterior distributions estimate well the "true" parameter values

used to generate y. Your OpenBUGS code will need to update the $n \times 1$ vector w from a multivariate normal distribution (note, OpenBUGS wants a precision matrix for this distribution, so you will need to invert the $\sigma^2 R(\phi)$ on each MCMC iteration). How do the runtimes compare between your OpenBUGS program and the spBayes model you fit in the exercise above?

References

Abdulah, S., Ltaief, H., Sun, Y., Genton, M. G., & Keyes, D. E. (2018). Exageostat: A high performance unified software for geostatistics on manycore systems. *IEEE Transactions on Parallel and Distributed Systems*, *29*(12), 2771–2784.

Bakar, K. S., & Sahu, S. K. (2018). *Spatio-temporal Bayesian modeling*. R package version 3.3.

Banerjee, S., Carlin, B. P., & Gelfand, A. E. (2014). *Hierarchical modeling and analysis for spatial data* (2nd ed.). Chapman & Hall/CRC monographs on statistics & applied probability. Taylor & Francis.

Banerjee, S. (2017). High-dimensional Bayesian geostatistics. *Bayesian Analysis*, *12*, 583–614.

Banerjee, S., Gelfand, A. E., Finley, A. O., & Sang, H. (2008). Gaussian predictive process models for large spatial data sets. *Journal of the Royal Statistical Society Series B*, *70*(4), 825–848.

Barbian, M. H., & Assunção, R. M. (2017). Spatial subsemble estimator for large geostatistical data. *Spatial Statistics*, *22*, 68–88.

Berliner, L. M. (1996). Hierarchical Bayesian time series models. In *Maximum Entropy and Bayesian Methods* (pp. 15–22). Springer.

Bivand, R. (2019). *CRAN task view: Analysis of spatial data*. Accessed 25 Feb 2019. https://cran.r-project.org/web/views/Spatial.html

Carlin, B. P., & Louis, T. A. (2009). *Bayesian methods for data analysis* (3rd ed.). Boca Raton: Chapman & Hall/CRC.

Chiles, J.-P., & Delfiner, P. (1999). *Geostatistics: modeling spatial uncertainty*. New York: Wiley.

Christensen, O. F., & Ribeiro, P. J. (2017). *A package for generalised linear spatial models*. R package version 0.9-11.

Cressie, N. (1993). *Statistics for spatial data*. Wiley series in probability and mathematical statistics: Applied probability and statistics. Hoboken: Wiley.

Cressie, N., & Wikle, C. K. (2011). *Statistics for spatio-temporal data*. CourseSmart series. Hoboken: Wiley. https://books.google.com/books?id=-kOC6D0DiNYC

Cressie, N., & Johannesson, G. (2008). Fixed rank kriging for very large spatial data sets. *Journal of the Royal Statistical Society B*, *70*, 209–226.

Datta, A., Banerjee, S., Finley, A. O., & Gelfand, A. E. (2016). Hierarchical nearest-neighbor Gaussian process models for large geostatistical datasets. *Journal of the American Statistical Association111*(514), 800–812. https://doi.org/10.1080/01621459.2015.1044091

Diggle, P., & Ribeiro, P. J. (2007). *Model-based geostatistics*. Springer series in statistics. New York: Springer. https://books.google.com/books?id=qCqOm39OuFUC

Ek, A. R. (1969). *Stem map data for three forest stands in northern Ontario*. Inf. Rep. O-X-113. Forest Research Laboratory, Sault Ste. Marie, Ontario.

Finley, A. O., & Banerjee, S. (2019). *Univariate and multivariate spatial-temporal modeling*. R package version 0.4.2.

Finley, A. O., & Banerjee, S. (2020). Bayesian spatially varying coefficient models in the spBayes R package. *Environmental Modelling & Software125*, 104608. http://www.sciencedirect.com/science/article/pii/S1364815219310412

Finley, A. O., Banerjee, S., & Carlin, B. P. (2007). spBayes: An R package for univariate and multivariate hierarchical point-referenced spatial models. *Journal of Statistical Software 19*(4), 1–24. http://www.jstatsoft.org/v19/i04/

Finley, A. O., Banerjee, S., & Gelfand, A. (2015). spBayes for large univariate and multivariate point-referenced spatio-temporal data models. *Journal of Statistical Software*. Articles *63*(13), 1–28. https://www.jstatsoft.org/v063/i13

Finley, A. O., Datta, A., & Banerjee, S. (2020). **spNNGP**: *Spatial regression models for large datasets using nearest neighbor Gaussian processes*. R package version 0.1.4. https://CRAN.R-project.org/package=spNNGP

Finley, A. O. (2011). Comparing spatially-varying coefficients models for analysis of ecological data with non-stationary and anisotropic residual dependence. *Methods in Ecology and Evolution, 2*(2), 143–154.

Finley, A. O., Sang, H., Banerjee, S., & Gelfand, A. E. (2009). Improving the performance of predictive process modeling for large datasets. *Computational Statistics & Data Analysis, 53*(8), 2873–2884.

Finley, A. O., Datta, A., Cook, B. D., Morton, D. C., Andersen, H. E., & Banerjee, S. (2019). Efficient algorithms for Bayesian nearest neighbor Gaussian processes. *Journal of Computational and Graphical Statistics, 28*, 401–414.

Furrer, R. (2016). **spam**: *SPArse matrix*. R package version 1.4-0. https://CRAN.R-project.org/package=spam

Furrer, R., & Sain, S. R. (2010). **spam**: A sparse matrix R package with emphasis on MCMC methods for Gaussian Markov random fields. *Journal of Statistical Software 36*(10), 1–25. http://www.jstatsoft.org/v36/i10/

Gelfand, A. E., Kim, H.-J., Sirmans, C. F., & Banerjee, S. (2003). Spatial modeling with spatially varying coefficient processes. *Journal of the American Statistical Association, 98*(462), 387–396.

Gelman, A., Carlin, J. B., Stern, H. B., Dunson, D. B., Vehtari, A., & Rubin, D. B. (2013). *Bayesian data analysis* (3rd ed.). New York: Chapman & Hall/CRC.

Gelman, A., Hwang, J., & Vehtari, A. (2014). Understanding predictive information criteria for Bayesian models. *Statistics and Computing, 24*, 997–1016.

Gerber, F. (2017). **gapfill**: *Fill missing values in satellite data*. R package version 0.9.5. https://CRAN.R-project.org/package=gapfill

Gneiting, T., & Raftery, A. E. (2007). Strictly proper scoring rules, prediction, and estimation. *Journal of the American Statistical Association 102*(477), 359–378. https://doi.org/10.1198/016214506000001437

Gramacy, R. B. (2016). **laGP**: Large-scale spatial modeling via local approximate Gaussian processes in R. *Journal of Statistical Software, 72*(1), 1–46.

Gramacy, R. B., & Apley, D. W. (2015). Local Gaussian process approximation for large computer experiments. *Journal of Computational and Graphical Statistics, 24*(2), 561–578.

Guhaniyogi, R., & Banerjee, S. (2018). Meta-kriging: Scalable Bayesian modeling and inference for massive spatial datasets. *Technometrics, 60*(4), 430–444.

Guinness, J. (2018). Permutation and grouping methods for sharpening Gaussian process approximations. *Technometrics, 60*(4), 415–429.

Guo, L., Ma, Z., & Zhang, L. (2008). Comparison of bandwidth selection in application of geographically weighted regression: A case study. *Canadian Journal of Forest Research, 38*(9), 2526–2534.

Heaton, M. J., Datta, A., Finley, A. O., Furrer, R., Guinness, J., Guhaniyogi, R., et al. (2019). A case study competition among methods for analyzing large spatial data. *Journal of Agricultural, Biological and Environmental Statistics, 24*(3), 398–425.

Katzfuss, M. (2017). A multi-resolution approximation for massive spatial datasets. *Journal of the American Statistical Association, 112*(517), 201–214.

Møller, J., & Waagepetersen, R. P. (2003). *Statistical inference and simulation for spatial point processes*. Chapman & Hall/CRC monographs on statistics & applied probability. Boca Raton: CRC Press.

Nychka, D., Bandyopadhyay, S., Hammerling, D., Lindgren, F., & Sain, S. (2015). A multiresolution Gaussian process model for the analysis of large spatial datasets. *Journal of Computational and Graphical Statistics*, 24(2), 579–599.

Ribeiro, P. J., & Diggle, P. J. (2018). *Analysis of geostatistical data*. R package version 1.7-5.2.1.

Risser, M. D., & Turek, D. (2019). Bayesian nonstationary Gaussian process modeling: The **BayesNSGP** package for R. arXiv preprint arXiv:1910.14101.

Rue, H., Martino, S., Lindgren, F., Simpson, D., Riebler, A., Krainski, E. T., & Fuglstad, G.-A. (2017). INLA: *Bayesian analysis of latent Gaussian models using integrated nested laplace approximations*. R package version 17.06.20. http://r-inla.org/

Saha, A. & Datta, A. (2018a). BRISC: Bootstrap for rapid inference on spatial covariances. *Stat* e184.

Saha, A., & Datta, A. (2018b). **BRISC**: *Fast inference for large spatial datasets using BRISC*. R package version 0.1.0. https://CRAN.R-project.org/package=BRISC

Sang, H., Jun, M., & Huang, J. Z. (2011). Covariance approximation for large multivariate spatial data sets with an application to multiple climate model errors. *The Annals of Applied Statistics* 2519–2548.

Schabenberger, O., & Gotway, C. A. (2004). *Statistical methods for spatial data analysis*. Chapman & Hall/CRC texts in statistical science. Boca Raton: Taylor & Francis. https://books.google.com/books?id=iVJuVLArmZcC

Stein, M. (1999). *Interpolation of spatial data: some theory for kriging*. Springer series in statistics. New York: Springer. https://books.google.com/books?id=5n_XuL2Wx1EC

Sun, Y., Li, B., & Genton, M. (2011). Geostatistics for large datasets. In J. Montero, E. Porcu, & M. Schlather (Eds.), *Advances and challenges in space-time modelling of natural events* (pp. 55–77). Berlin: Springer.

Tobler, W. R. (1970). A computer movie simulating urban growth in the Detroit region. *Economic Geography 46*(sup1), 234–240. https://www.tandfonline.com/doi/abs/10.2307/143141

Wackernagel, H. (2003). *Multivariate geostatistics: An introduction with applications*. Berlin: Springer. https://books.google.com/books?id=Rhr7bgLWxx4C

Zammit-Mangion, A., & Cressie, N. (2017). **FRK**: An R package for spatial and spatio-temporal prediction with large datasets. arXiv preprint arXiv:1705.08105.

Appendix A
Some Common Conjugate Models

Consult Chap. 2 for more information on densities, including parameter ranges, etc.

1. Beta-Binomial model

   - Likelihood: $y \mid p \sim \mathbf{Bi}(n, p)$
   - Prior density: $p \sim \mathbf{Be}(\alpha, \beta)$
   - Posterior density: $p \mid y \sim \mathbf{Be}(\alpha + y, n - y + \beta)$

2. Dirichlet-Multinomial model

   - Likelihood: $\boldsymbol{y} \mid \boldsymbol{p} \sim \mathbf{Mu}(n, \boldsymbol{p})$
   - Prior density: $\boldsymbol{p} \sim \mathbf{Dir}(\boldsymbol{\alpha})$
   - Posterior density: $\boldsymbol{p} \mid \boldsymbol{y} \sim \mathbf{Dir}(\boldsymbol{y} + \boldsymbol{\alpha})$

3. Gamma-Poisson model

   - Likelihood: $y \mid \lambda \sim \mathbf{Poi}(\lambda)$
   - Prior density: $\lambda \sim \mathbf{Ga}(\alpha, \beta)$
   - Posterior density: $\lambda \mid y \sim \mathbf{Ga}(\sum_{i=1}^{n} y_i + \alpha, n + \beta)$

4. Normal-Normal model, known Normal variance

   - Likelihood: $y \mid \mu \sim \mathbf{N}(\mu, \sigma^2)$
   - Prior density: $\mu \sim \mathbf{N}(\theta, \tau^2)$
   - Posterior density: $\mu \mid y \sim \mathbf{N}(\delta, \nu), \ \delta = \frac{y\tau^2 + \theta\sigma^2}{\sigma^2 + \tau^2}, \ \nu = \frac{\sigma^2\tau^2}{\sigma^2 + \tau^2}$

5. Inverse Gamma-Normal model, known Normal mean

   - Likelihood: $y \mid \sigma^2 \sim \mathbf{N}(\mu, \sigma^2)$
   - Prior density: $\sigma^2 \sim \mathbf{Ga}^{-1}(\alpha, \beta)$
   - Posterior density: $\sigma^2 \mid y \sim \mathbf{Ga}^{-1}(\mu + \alpha, y + \beta)$

For the Normal model with unknown mean and unknown variance, see Sect. 4.3.

© Springer Nature Switzerland AG 2020
E. J. Green et al., *Introduction to Bayesian Methods in Ecology and Natural Resources*,
https://doi.org/10.1007/978-3-030-60750-0

Appendix B
Markov Chain Monte Carlo Sampling

Markov chain Monte Carlo (MCMC) sampling is at the heart of modern Bayesian inference. MCMC methods gained wide notoriety following publication of the landmark papers by Casella and George (1992), Gelfand (2000). In this section we briefly explain one MCMC method: Gibbs sampling. For more information, readers are directed to Gelfand and Smith (1990), Gelfand et al. (1990), Gilks and Wild (1992), among others. For an accessible introduction to sampling-resampling methods and an illustration of how data is filtered through a prior distribution to form a posterior distribution, readers are directed to Lunn et al. (2013).

In this section, we adopt the notation of Gelfand (2000). We will denote the distribution of U by $[U]$, the joint distribution of U and V by $[U, V]$, and the conditional distribution of U and V, given W, by $[U, V \mid W]$.

Suppose our target distribution is the joint distribution $[U, V, W]$ but we cannot solve for it analytically. However suppose that we *can* generate samples from $[U \mid V, W]$, $[V \mid U, W]$, and $[W \mid U, V]$. These distributions are called the *full conditional* distributions of U, V, and W because the conditional distribution of each quantity is conditioned on *all* the other quantities.

Let $U^{[0]}$, $V^{[0]}$, and $W^{[0]}$ be initial guesses (or starting values) for U, V, and W. Then we proceed as follows:

1. Generate $U^{[1]}$ from $[U \mid V^{[0]}, W^{[0]}]$.
2. Generate $V^{[1]}$ from $[V \mid U^{[1]}, W^{[0]}]$.
3. Generate $W^{[1]}$ from $[W \mid U^{[1]}, V^{[1]}]$.

Iterate steps 1–3 a large number of times, incrementing each bracketed superscript by 1 on each iteration.

Under very mild regularity conditions, after a sufficient number of "burn-in" iterations of these steps, the generated values for U, V, and W will be samples from the target joint distribution $[U, V, W]$. As Gelfand and Smith (1990) mention, there is no other place for the joint density to go.

How does the above help us? Suppose U, V, and W are the parameters of our sampling model and all the distributions in steps 1–3 are *also* conditioned on the data (D). Then iteratively sampling from $[U \mid V, W, D]$, $[V \mid U, W, D]$, and $[W \mid U, V, D]$ will

© Springer Nature Switzerland AG 2020
E. J. Green et al., *Introduction to Bayesian Methods in Ecology and Natural Resources*,
https://doi.org/10.1007/978-3-030-60750-0

generate samples from $[U, V, W \mid D]$. The latter is the joint posterior distribution, the exact distribution we require for Bayesian inference!

In order to complete the three steps above, we need to be able to generate samples from each of the full conditional distributions. Sometimes this is possible and sometimes it isn't. When it's not, we can substitute a Metropolis sampling step for whichever full conditional we cannot sample from (e.g., see Robert and Casella 1999). This was formerly called "Metropolis within Gibbs" sampling but now is simply referred to as MCMC sampling. One improvement over Metropolis within Gibbs is to adaptively modify what is called the proposal distribution in the Metropolis step as the sampler runs. The latter is called "adaptive-rejection" sampling (Smith and Gelfand 1992) and results in a more efficient algorithm.

Access to samples from an arbitrarily large joint posterior sample allows us to take advantage of the duality between samples and densities. We can learn about any feature of the joint posterior distribution by studying that feature in the joint posterior sample. If we are interested in the marginal posterior sample of a particular parameter, say U, then we just focus on the values of U in the joint posterior sample. Taken by themselves, these values are a sample from the marginal posterior sample for U.

Suppose we are interested in the posterior mean of U. We can estimate it with the mean of the marginal posterior sample for U. Also, we can form approximate $(1-\alpha)\%$ credible intervals by computing the $\alpha/2$ and $1 - \alpha/2$ quantiles of U from its marginal posterior sample.

Finally, we can derive the posterior distribution of any function of the parameters by computing the function for each observation in the joint posterior sample. For instance, suppose we are interested in the function $(U - V)$. We can derive the marginal posterior sample of the function by computing $u_i - v_i$ $i = 1, 2, \ldots, K$, where K is the size of the joint posterior sample.

Appendix C
Short Tutorial on **OpenBUGS**

There are many software packages available for Bayesian analysis, such as OpenBUGS, JAGS, Stan, and the R packages **arm, bayesm, LaplacesDemon, MCMCpack, mcmc** and **nimble**. Our reliance on OpenBUGS is not meant as an endorsement of it. Any of the available Bayesian software packages available can be used to accomplish the goal of fitting Bayesian models. We chose to emphasize OpenBUGS because it is the descendant of WinBUGS and, before that, BUGS, the original software package for MCMC sampling for Bayesian models. Hence we believe that more scientists are familiar with it. It is our belief that once a scientist is comfortable with programming in any of the software packages available, conversion to a different package is *relatively* painless.

OpenBUGS can be used as a stand-alone package or called from R using **R2OpenBUGS**. We have chosen to present our OpenBUGS code as stand-alone. This stems from our experience teaching this material to graduate students for over a decade. However, once a scientist has developed some competence and experience in OpenBUGS programming, it is easy to use **R2OpenBUGS** in lieu of the OpenBUGS environment. In fact, in our actual work, we are much more likely to use **R2OpenBUGS** because doing so affords us ready access to the powerful data manipulation capabilities of R. In either case, the key step is the description of the model in the OpenBUGS language, and this is the same whether it is done in the OpenBUGS environment or stored in a separate .txt file and called using the "model.file=" statement in **R2OpenBUGS**.

Windows, Mac and Linux versions of OpenBUGS can be downloaded and installed from http://www.openbugs.net/w/Downloads. Note: as of April, 2020, the Mac version uses WINE. Users must check to insure that their OS is compatible with WINE. If not, a potential solution is to use a commercial package to emulate a Windows OS and install the Windows version of OpenBUGS.

Starting OpenBUGS puts the user into the OpenBUGS environment, as displayed in Fig. C.1. First-time users should click on "Manuals" and then "OpenBUGS User Manual" and read the latter. The manual contains much more information on the OpenBUGS language than is presented here. A further excellent source on

© Springer Nature Switzerland AG 2020

E. J. Green et al., *Introduction to Bayesian Methods in Ecology and Natural Resources*,
https://doi.org/10.1007/978-3-030-60750-0

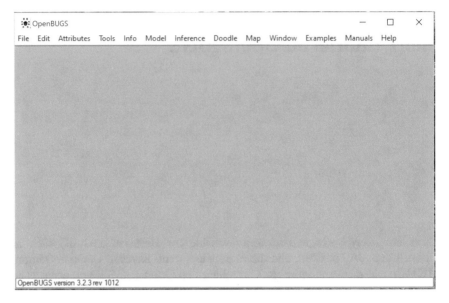

Fig. C.1 OpneBUGS start-up screen

OpenBUGS is Tierney (1994), a text authored by the OpenBUGS development team.

A standard program in OpenBUGS consists of three steps: (1) a model step which describes the model to be fitted, (2) a data step in which the data are described and input, and (3) an initial values step in which stochastic nodes are given starting values. We will present some background on each step in turn.

C.1 Model Step

The model step is where the model is described. OpenBUGS recognizes two type of objects: stochastic nodes and deterministic nodes. Stochastic nodes appear on the left-hand side of the "∼" operator which should be read as "is distributed as". Deterministic nodes appear on the left-hand side of the "<-" operator which should be read as "is replaced by." An argument can only appear on the left-hand side of an expression once in an OpenBUGS program. Looping, for instance over all the observations in a data set, is available using the `for{}` statement.

A large number of functions and distributions are available in OpenBUGS, and these are listed in the User Manual. In the event that a user needs access to a distribution which is not automatically available, it can be defined using the generic `dloglike` distribution, as described in the User Manual. When using an available distribution, users should **always** check the definition of the distribution in the User

Manual, as many distributions can be written multiple ways and it is necessary to ensure the definition OpenBUGS uses is the one the user expects. In particular, OpenBUGS defines the commonly used Normal distribution in terms of its mean and *precision* rather than the more common mean and variance. In the authors' experience, specifying the normal distribution incorrectly is the most common error in OpenBUGS programming.

C.2 Data Step

This is the step in which the data are entered. There are a number of formats for entering the data, each of which is described in the User Manual, under "Model Specification." For small data sets, we find the R\Splus list format the most convenient, whereas for larger data sets we prefer the rectangular format. In the latter case, we often find it easiest to have the data in a separate window than the model.

C.3 Initial Values Step

Here is where initial values are supplied for each stochastic node. Although Open-BUGS can generate initial values from the prior distributions for stochastic nodes, it is generally preferable for the user to supply them. For nodes with vague prior distributions, OpenBUGS might generate wildly inappropriate missing values, causing the MCMC sampler to require many more iterations to converge, or to encounter numerical instability difficulties.

C.4 A Few OpenBUGS "Tricks"

The user must specify any parameter they wish to monitor in the Sample Monitor tool, under the Inference tab. Then, the joint posterior sample of any or all of these parameters is available under the coda option in the Sample Monitor tool. If a user wants access to the joint posterior sample of *all* the parameters specified in the Sample Monitor tool, it is only necessary to enter * in the node window.[1]

Requesting the joint posterior sample with the coda opens two windows: (1) an output file and (2) an index file describing the format of the output file. If the user intends to use the R package **coda** to analyze to output for convergence or to produce estimates or plots of posterior densities, then the index and output files must both be

---

[1]One requests the coda output *after* the sample is generated in OpenBUGS, but one must specify the nodes to monitor *before* generating the sample.

stored as `.txt` files. However, the suffix on the index filename must be `.ind` and the suffix on the output filename must be `.out`.

Suppose we wish to store the index and output files as `.txt` under the names `test.ind` and `test.out`. First highlight the index file. Then under the `File` tab on the far left of the OpenBUGS menu bar, select `Save As...`. This opens up a window with spaces to enter the filename and file type. The file type option has a drop-down menu. In the drop-down menu, select `.txt`. Next in the file name space, enter `"test.ind"`, complete with the quotation marks. This will save the file as `text.ind`, but it will be in a `.txt` format. Do the same with the output file, changing `test.ind` to `test.out`.

Finally, a useful modeling "trick" is nested indexing. Suppose the data arise in m groups, and we wish to fit the model

$$y_{ij} \sim \mathbf{N}(\mu_i, \sigma^2), \ i = 1, 2, \ldots, m, \ j = 1, 2, \ldots, n_i. \qquad \text{(C.4.1)}$$

For the moment, we won't concern ourselves with the prior distributions on μ_i and σ^2.

Next suppose we include the variable `grp` in the data, indicating which group each observation belongs to, i.e., for each observation, `grp` takes on one of the values in the set $(1,2,\ldots,m)$, and that there are N total observations in the data. Then the following coding can be useful:

```
for (i in 1:N){
    y[i] <- dnorm(mu[grp[i]], tau)
}
```

Of course, in the above `tau` is the precision, i.e., the inverse of the variance. But our interest here is on `mu[grp[i]]`. The variable `grp` will be read for each observation in the data and `grp[i]` will return a value of 1, 2, …, or m, allowing the observation y_i to be drawn from the normal distribution for the correct group.

C.5   Convergence

As mentioned above, OpenBUGS can generate samples in a format designed for use in the R package **coda**. The latter includes four common tests for convergence of the sampler. Readers familiar with classical procedures such as nonlinear least squares may be familiar with the notion of convergence to a point, typically the minimum or maximum of a function. Convergence in MCMC sampling is different. In the latter situation we are assessing when the sampler has reached a stage such that all subsequent sample values can be regarded as arising from the target joint posterior distribution.

While scientists certainly ought to monitor MCMC samplers for convergence, there is less of an emphasis on testing for convergence in modern Bayesian analysis than there was at the beginning of the MCMC revolution. This is due to the advances in hardware. Whereas it was once usual to run MCMC samplers for $\sim$1000 iterations, it is now commonplace to run them for 100,000 or more. Given the ease of generating enormous posterior samples, we find it sufficient to monitor the history and quantiles (say the 0.025 and 0.975 quantiles) of the parameters. When these are steady for a large number of iterations, we are confident that the sampler has converged. Having said that, we do recommend that users run multiple chains from varying initial values to ensure that they all converge to the same parameter space (if they don't, this is evidence of a lack of convergence). Once the user determines the number of iterations required to attain convergence, then it would be pragmatic to run one final "production" chain for 5–10 times more iterations than are requited for convergence, and base all inference on the production chain.

References

Casella, G., & George, E. I. (1992). Explaining the Gibbs sampler. *The American Statistician, 46*(3), 167–174.

Gelfand, A. E. (2000). Gibbs sampling. *Journal of the American Statistical Association, 95*(452), 1300–1304.

Gelfand, A. E., & Smith, A. F. M. (1990). Sampling-based approaches to calculating marginal densities. *Journal of the American Statistical Association, 85*(410), 398–409.

Gelfand, A. E., Hills, S. E., Racine-Poon, A., & Smith, A. F. M. (1990). Illustration of Bayesian inference in normal data models using Gibbs sampling. *Journal of the American Statistical Association, 85*(412), 972–985.

Gilks, W. R., & Wild, P. (1992). Adaptive rejection sampling for Gibbs sampling. *Applied Statistics, 41*(2), 337–348.

Lunn, D., Jackson, C., Best, N., Thomas, A., & Spiegelhalter, D. (2013). *The BUGS book: A practical introduction to Bayesian analysis*. Boca Raton: CRC Press.

Robert, C., & Casella, G. (1999). *Monte Carlo statistical methods*. New York: Springer.

Smith, A. F. M., & Gelfand, A. E. (1992). Bayesian statistics without tears: A sampling-resampling perspective. *The American Statistician, 46*(2), 84–88.

Tierney, L. (1994). Markov chains for exploring posterior distributions. *The Annals of Statistics, 22*(4), 1701–1762.

© Springer Nature Switzerland AG 2020
E. J. Green et al., *Introduction to Bayesian Methods in Ecology and Natural Resources*,
https://doi.org/10.1007/978-3-030-60750-0

Printed in the United States
by Baker & Taylor Publisher Services